The ideas of particle physics

The ideas of particle physics

AN INTRODUCTION FOR SCIENTISTS

J.E. Dodd

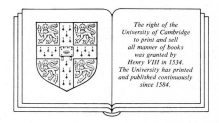

The right of the University of Cambridge to print and sell all manner of books was granted by Henry VIII in 1534. The University has printed and published continuously since 1584.

Cambridge University Press

CAMBRIDGE

LONDON   NEW YORK   NEW ROCHELLE

MELBOURNE   SYDNEY

7208-4716

PHYSICS

Published by the Press Syndicate of the University of Cambridge
The Pitt Building, Trumpington Street, Cambridge CB2 1RP
32 East 57th Street, New York, NY 10022, USA
296 Beaconsfield Parade, Middle Park, Melbourne 3206, Australia

First published 1984

Printed in Great Britain by
The Alden Press, Oxford

Library of Congress catalogue card number: 83-15227

British Library cataloguing in publication data

Dodd, James E.
The ideas of particle physics.

1. Particles (Nuclear physics)
I. Title
539.7′21   QC 793.2

ISBN 0 521 25338 1 hard covers
ISBN 0 521 27322 6 paperback

AL

Contents

To my parents

Preface

The last decade has seen an enormous advance in our understanding of the microworld. We now believe that we have a very firm idea of the basic material particles from which all observable matter is made. Similarly, of the four separately identifiable forces of nature, we believe we have a good unified theory of two of them, with the likelihood of a third being accommodated soon. Even the fourth, inscrutable gravity, is seemingly amenable to some of the more speculative theories.

In the late 1960s, it was a very different picture. The newly constructed high-energy accelerators produced hundreds of elementary particles as if to astonish physicists with the fecundity of matter beyond the atoms of the everyday world. Of the four forces of nature only one, the electromagnetic force, was properly understood.

Then, in rapid succession, a theoretical breakthrough led to the development of a common 'gauge theory' approach to three of the forces of nature – which was soon supported in 1973 by the discovery of the 'weak neutral currents'. The very next year the rather unexpected discovery of the psi meson and its close relatives lent convincing support for the physical reality of quarks, the hidden building blocks of the multitude of particles discovered previously. Since then, further theoretical advances and experimental discoveries have reinforced our acceptance of the nature of the elementary particles and the theories which we must use to describe them.

In many ways, the story is reminiscent of that of the 1920s when the experiments of atomic physics and the development of quantum theory led to a satisfactory comprehension of matter on the atomic level (i.e. involving distances of 10^{-8} cm and energies of a few electron volts). The story of the last decade is concerned with distances down to 10^{-16} cm and with energies of up to a few hundred billion electron volts, (10^{11} eV).

This book aims to explain, to a scientific audience, the story of the microworld over the last decade. The way of telling the story is largely chronological, but with reversion to logical subject divisions where appropriate. For instance, the advances described in Parts 5, 6, 7, and 8 were happening simultaneously in a complex interrelated fashion over the last decade or so. Although the main logical connections are mentioned, no attempt has been made to keep an accurate history of the development.

Part 0 takes up the story at the turn of the century when physics was first beginning to glimpse the remarkable nature of ordinary matter. Essential elements for understanding the microworld are introduced next: the theories of special relativity and quantum mechanics. These are the unshakeable pillars on which all of our story is subsequently founded. Immediately the two are combined in relativistic quantum theory which is the inescapable language of modern microphysics.

Part 1 examines, in turn, each of the four separate forces currently believed to govern all observed phenomena in the known universe, and goes on to describe the beginnings of particle physics proper with the discoveries of the muon (in 1937) and the pion (in 1947).

The story then progresses on to high-energy hadron physics (physics related to the pion and the proton), which dominated the world of particle physics in the 1960s. This was the logical extension of nuclear physics but it did not reveal any new fundamental insights (apart from the plethora of hadrons).

Parts 3 and 4 describe the physics of the weak nuclear force which had been evolving from the study of nuclear β decay since the 1930s. The discovery of parity violation in the weak interaction in the mid 1950s provided the impetus for a better theoretical understanding and this led eventually to the modern gauge theory of the weak force, which is described in Part 5. This is the first of the essentially modern topics described in the book.

Around the same time as the development of the weak interaction gauge theory, the so-called deep inelastic scattering experiments began to probe the interior of the proton with very high-energy photons. These experiments, which are described in Part 6, provided a first indication of point-like objects within the proton and so led to the idea that such objects (called quarks) may be more than just mathematical constructs required to lend some order to the patterns of the multitudinous hadrons.

As the physical reality of quarks became more accepted, a theory was formulated in an attempt to explain their behaviour. The splendidly named 'quantum chromodynamics', described in Part 7, offers a good description of the quarks' behaviour in some circumstances and benefits from sharing the same method of formulation as the earlier weak interaction gauge theory. The best class of reactions in which to observe the behaviour of quarks, and thus to test quantum chromodynamics, is electron–positron annihilation and this is described in Part 8. Over the last decade these reactions have allowed the discovery of two new types of quark and one new lepton (a heavier relation of the electron and the muon) and have provided firm evidence for the validity of quantum chromodynamics.

Finally, the last part of the book describes the most recent triumph of weak interaction gauge theory, the current experimental discovery of the class of particles responsible for mediating the force. Part 9 then goes on to describe various theories which attempt to unify the description of the various forces of nature within the framework of a 'grand (or super) unified gauge theory'.

The book has attempted to provide a fairly comprehensive view of the subject within a, hopefully, concise scope by assuming, on the part of the reader, a level of sophistication not usually required for popular texts. It is pitched at the level of graduates in the physical sciences, mathematics or other numerate subjects. For although no mathematical derivations are used and no complex formulae are given (other than by way of illustration), the text does assume a familiarity with basic physical concepts (mass, momentum, energy etc.) and does introduce concepts such as abstract spaces and group theory.

One or two words of caution on what the book is not are also appropriate. It is not a text book and makes no claim to the accuracy required in such a book or to the suitability for courses that a text book must incorporate. The diagrams in this book are for the purposes of illustration only and not for the purpose of conveying data. Also, the mathematical expressions are used not as working formulae but only as symbolic summaries of logical relationships described in the text. For instance, the overall tone of the book should be taken as $\approx$ instead of a definitive $=$.

During the somewhat lengthy gestation of the book, I have enjoyed pleasant and informative comments from Dr Ian Aitchison, Mr John Kleeman, Dr Ron Larham, Dr Jack Paton and Dr Peter Scharbach. Also, the referees of Cambridge University Press have provided many valuable thoughts on portions of the text. To all of the above I am most grateful for their time and forebearance, as well as for the variety and validity of their advice. Needless to say, any odd error or inaccuracy outstanding is my very own (if not actual policy). Finally, I would like to thank both Barbara Whiteside and Kate Mumford for their extremely efficient word-processing of the original manuscript.

James Dodd
London
May 1983

Part 0
Introduction

1
Matter and light

1.1 Introduction

The physical world we see around us has two main components: matter and light, and it is the modern explanation of these things which is the purpose of this book. During the course of the story, these two concerns may be restated as those of material particles and the forces which act between them, and we will most assuredly encounter new and exotic forms of both particles and forces. But in case we become distracted and confused by the elaborate and almost wholly alien contents of the microworld, let us remember that the origin of the story, and the motivation for all that follows, is the explanation of everyday matter and visible light.

Beginning as it does, with a laudable sense of history, at the turn of the century, the story is one of twentieth-century achievement. For the background, we have only to appreciate the level of understanding of matter and light around 1900, and some of the problems in this understanding, to prepare ourselves for the story of progress which follows.

1.2 The nature of matter

By 1900 most scientists were convinced that all matter is made up of a number of different sorts of atoms, as had been conjectured by the ancient Greeks millenia before and as had been indicated by chemistry experiments over the preceding two centuries. In the atomic picture, the different types of substance can be seen as arising from different arrangements of the atoms. In solids, the atoms are relatively immobile and in the case of crystals are arranged in set patterns of impressive precision. In liquids they roll loosely over one another and in gases they are widely separated and fly about at a velocity depending on the temperature of the gas, see Figure 1.1. The application of heat to a substance can cause phase transitions in which the atoms change their mode of behaviour as the heat energy is transferred into the kinetic energy of the atoms' motions.

Many familiar substances consist not of single atoms, but of definite combinations of certain atoms called molecules. In such cases it is these molecules which behave in the manner appropriate to the type of substance concerned. For instance, water consists of molecules, each made up of two hydrogen atoms and one oxygen atom. It is the molecules which are subject to a specific static arrangement in solid ice, the molecules which roll over each other in water and the molecules which fly about in steam.

The laws of chemistry, most of which were

Fig. 1.1. (*a*) Static atoms arranged in a crystal. (*b*) Atoms rolling around in a liquid. (*c*) Atoms flying about in a gas.

(*a*)

(*b*)

(*c*)

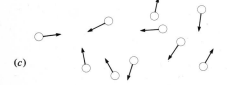

discovered empirically between 1700 and 1900, contain many deductions concerning the behaviour of atoms and molecules. At the risk of brutal over-simplification the most important of these can be summarised as follows:

(1) Atoms can combine to form molecules, as indicated by chemical elements combining only in certain proportions (Richter and Dalton).

(2) At a given temperature and pressure, equal volumes of gas contain equal numbers of molecules (Avogadro).

(3) The relative weights of the atoms are approximately integral multiples of the weight of the hydrogen atom (Prout).

(4) The mass of each atom is associated with a specific quantity of electrical charge (Faraday and Webber).

(5) The elements can be arranged in families having common chemical properties but different atomic weights (Mendeleeff's periodic table).

(6) An atom is approximately 10^{-10} m across, as implied by the internal friction of a gas (Loschmidt).

One of the philosophical motivations behind the atomic theory (a motivation we shall see repeated later), was the desire to explain the diversity of matter by assuming the existence of just a few fundamental and indivisible atoms. But by 1900 over 90 varieties of atoms were known, an uncomfortably large number for a supposedly fundamental entity. Also there was evidence for the disintegration (divisibility) of atoms. At this breakdown of the 'ancient' atomic theory, modern physics begins.

1.3   Atomic radiations
1.3.1   *Electrons*

In the late 1890s, J. J. Thompson of the Cavendish Laboratory at Cambridge was conducting experiments to examine the behaviour of gas in a glass tube when an electric field was applied across it. He came to the conclusion that the tube contained a cloud of minute particles with negative electrical charge – the electrons. As the tube had been filled only with ordinary gas atoms, Thompson was forced to conclude that the electrons had originated within the supposedly indivisible atoms. As the atom as a whole is electrically neutral, on releasing a negatively charged electron the remaining part, the ion, must carry the equal and opposite positive charge. This was entirely in accord with the long-known results of Faraday's electrolysis experiments, which required a specific electrical charge to be associated with the atomic mass.

By 1897, Thompson had measured the ratio of the charge to the mass of the electron (denoted e/m) by observing its behaviour in magnetic fields. By comparing this number with that of the ion, he was able to conclude that the electron is thousands of times less massive than the atom (and some 1837 times lighter than the lightest atom, hydrogen). This led Thompson to propose his 'plum-pudding' picture of the atom, in which the small negatively charged electrons were thought to be dotted in the massive, positively charged body of the atom, see Figure 1.2.

1.3.2 *X-rays*

At about the time of Thompson's work, the German Wilhelm Röntgen used electrons to discover a new form of penetrating radiation. This radiation was emitted when a stream of fast electrons struck solid matter and were thus rapidly deaccelerated. This was achieved by boiling the electrons out of a metallic electrode in a vacuum tube and accelerating them into another electrode by applying an electric field across the two, see Figure 1.3. Very soon the X-rays were identified as another form of electromagnetic radiation, i.e. radiation that is basically the same as visible light, but with a much higher frequency and shorter

wavelength. An impressive demonstration of the wave nature of X-rays was provided in 1912 when the German physicist, Max von Laue shone them through a crystal structure. In doing so, he noticed the regular geometrical patterns characteristic of the diffraction which occurs when a wave passes through a regular structure whose characteristic size is comparable to the wavelength of the wave. In this case, the regular spacing of atoms within the crystal is about the same as the wavelength of the X-rays. Although these X-rays do not originate from within the structure of matter, we shall see next how they are the close relatives of radiations which do.

1.3.3 *Radioactivity*

At the same time as the work taking place on electrons and X-rays, the French physicist Becquerel and Pierre and Marie Curie were conducting experiments on the heavy elements. During his study of uranium salts, Becquerel noticed the emission of radiation rather like that which Röntgen had discovered. But Becquerel was doing nothing to his uranium: the radiation was emerging spontaneously. Similarly, by 1898, the Curies had discovered that the element radium also emits copious amounts of radiation.

These early experimenters first discovered the radiation through its darkening effect on photo-

Fig. 1.3. The production of X-rays by colliding fast electrons with matter.

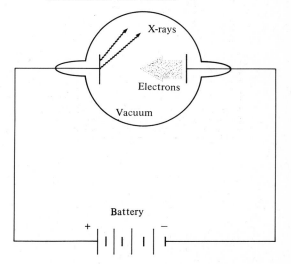

Fig. 1.2. Thompson's 'plum-pudding' picture of the atom.

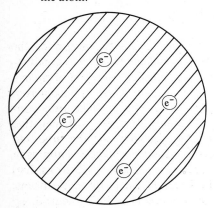

graphic plates, and this became the standard early method for detecting radiation: by developing photographic plates left in its path. Physicists had to rely on this rather tedious method until 1912 when C.T.R. Wilson of the Cavendish invented the cloud chamber. This device encourages easily visible water droplets to form around the atoms, which have been ionised (have had an electron removed) by the passage of the radiation through air. This provides a plan view of the path of the radiation and so gives us a clear picture of what is happening. (Sophisticated variants of the device are still in use today as the massive bubble chambers in high-energy accelerator experiments.)

If a radioactive source such as radium is brought close to the cloud chamber, the emitted radiation will trace paths in the chamber. When a magnetic field is placed across the chamber, then the radiation paths will separate into three components which are characteristic of the type of radiation, see Figure 1.4. The first component of radiation (denoted α) is bent slightly by the magnetic field, which indicates that the radiation carries electric charge. Measuring the radius of curvature of the path in a given magnetic field can tell us that it is made up of massive particles with two positive electric charges. These particles can be identified as the nuclei of helium atoms, often referred to as α

Fig. 1.4. Three components of radioactivity displayed in a cloud chamber. ⊙ signifies that the direction of the applied magnetic field is perpendicular to, and out of the plane of, the paper.

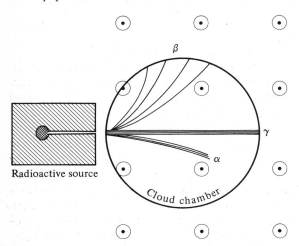

particles. Furthermore, these α particles always seem to travel a fixed distance before being stopped by collisions with the air molecules. This suggests that they are liberated from the source with a constant amount of energy and that the same internal reactions within the source atoms are responsible for all α particles.

The second component of the radiation (denoted γ) is not at all affected by the magnetic field, showing that it carries no electric charge, and it is not stopped by collisions with the air molecules. These γ-rays were soon identified as the close relatives of Röntgen's X-rays but with even higher frequencies and even shorter wavelengths. The γ-rays can penetrate many centimetres of lead before being absorbed. They are the products of reactions occurring spontaneously within the source atoms, which liberate large amounts of electromagnetic energy but no material particles, indicating a different sort of reaction to that responsible for α-rays.

The third component (denoted β radiation) is bent significantly in the magnetic field in the opposite direction to the α-rays. This is interpreted as single, negative electrical charges with much lesser mass than the α-rays. They were soon identified as the same electrons as those discovered by J. J. Thompson, being emitted from the source atoms with a range of different energies. The reactions responsible form a third class distinct from the origins of α- or γ-rays.

The three varieties of radioactivity have a double importance in our story. Firstly, they result from the three main fundamental forces of nature effective within atoms. Thus the phenomenon of radioactivity may be seen as the cradle for all of what follows. Secondly, and more practically, it was the products of radioactivity which first allowed physicists to explore the interior of atoms and which later indicated totally novel forms of matter, as we will see in due course.

1.4 Rutherford's Atom

In the first decade of the twentieth century, Rutherford had pioneered the use of naturally occurring atomic radiations as probes of the internal structure of atoms. In 1911, at the Cavendish Laboratory, he suggested to his colleagues, Geiger and Marsden, that they allow the α particles emitted from a radioactive element to pass through a thin gold foil and observe the deflection of the outgoing α particles from their original path, see Figure 1.5. On the basis of Thompson's 'plum-pudding' model of the atom, they should experience only slight deflections, as nowhere in the uniformly occupied body of the atom would the electric field be enormously high. But the experimenters were surprised to find that the heavy α particles were sometimes drastically deflected, occasionally bouncing right back towards the source. In a dramatic analogy attributed (somewhat dubiously) to Rutherford: 'It was almost as incredible as if you fired a 15-inch shell at a piece of tissue paper and it came back and hit you!'

The implication of this observation is that a very strong repulsive force must be at work within the atom. This force cannot be due to the electrons as they are over 7000 times lighter than the α particles and so can exert only minute effects on the α-particle trajectories. The only satisfactory explanation of the experiment is that all the positive electric charge in the atom is concentrated in a small nucleus at the middle, with the electrons orbiting the nucleus at some distance. By assuming that the entire positive charge of the atom is concentrated

Fig. 1.5. The Geiger and Marsden experiment. According to Rutherford's scattering formula, the number of α particles scattered through a given angle decreases as the angle increases away from the forward direction.

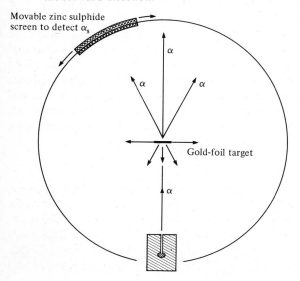

Movable zinc sulphide screen to detect α_s

α

α   α

Gold-foil target

α

α

with the atomic mass in a small central nucleus, Rutherford was able to derive his famous scattering formula which describes the relative numbers of α particles scattered through given angles on colliding with an atom, see Figure 1.5.

Rutherford's picture of the orbital atom is in contrast with our perception of apparently 'solid' matter. From the experiments he was able to deduce that the atomic nucleus, which contains 99.9% of the mass of the atom, has a diameter of about 10^{-15} m compared to an atomic diameter of about 10^{-10} m. For illustration, if we took a cricket ball to act as the nucleus, the atomic electrons would be ping pong balls 5 km distant! Such an analogy brings home forcibly just how sparse apparently solid matter is and just how dense is the nucleus itself. But despite this clear picture of the atom, indicated from the experiment, explaining how it works is fraught with difficulties, as we shall see in Chapter 3.

1.5   Two problems

Just as these early atomic experiments revealed an unexpected richness in the structure of matter, so too, theoretical problems forced upon physicists more-sophisticated descriptions of the natural world. The theories of special relatively and quantum mechanics arose as physicists realised that the classical physics of mechanics, thermodynamics and electromagnetism were inadequate to account for apparent mysteries in the behaviour of matter and light. Historically, the mysteries were contained in two problems, both under active investigation at the turn of the century.

1.5.1   *The constancy of the speed of light*

Despite many attempts to detect an effect, no variation was discovered in the speed of light. Light emerging from a torch at rest seems to travel forward at the same speed as light from a torch travelling at arbitrarily high speeds. This is very different from the way we perceive the behaviour of velocities in the everyday world. But, of course, we humans never perceive the velocity of light, it is just too fast! This unexpected behaviour is not contrary to common experience, it is beyond it. Explanation for the behaviour forms the starting point for the theory of special relativity, which is the necessary description of anything moving very fast (i.e. nearly all elementary particles), see Chapter 2.

1.5.2 *The interaction of light with matter*

All light, for instance sunlight, is a form of heat and so the description of the emission and absorption of radiation from matter was approached as a thermodynamical problem. In 1900 the German physicist, Max Planck, concluded that the classical thermodynamical theory was inadequate to describe the process correctly. The classical theory seemed to imply that if light of any one colour (any one wavelength) could be emitted from matter in a continuous range of energy down to zero, then the total amount of energy radiated by the matter would be infinite. Much against his inclination, Planck was forced to conclude that light of any given colour cannot be emitted in a continuous band of energy down to zero, but only in multiples of a fundamental quantum of energy, representing the minimum negotiable bundle of energy at any particular wavelength. This is the starting point of quantum mechanics, which is the necessary description of anything very small (i.e. all atoms and elementary particles), see Chapter 3.

As the elementary particles are both fast moving and small, it follows that their description must incorporate the rules of both special relativity and quantum mechanics. The synthesis of the two is known as relativistic quantum theory and this is described briefly in Chapter 4.

2

Special relativity

2.1 Introduction

A principle of relativity is simply a statement reconciling the points of view of observers who may be in different physical situations. Classical physics relies on the Galilean principle of relativity, which is perfectly adequate to reconcile the points of views of human observers in everyday situations. But modern physics requires the adoption of Einstein's special theory of relativity, as it is this theory which is known to account for the behaviour of physical laws when very high velocities are involved (typically those at or near the speed of light, denoted c).

It is an astonishing tribute to Einstein's genius that he was able to infer the special theory of relativity in the almost total absence of the experimental evidence which is now commonplace. He was able to construct the theory from the most tenuous scraps of evidence.

To us lesser mortals, it is challenge enough to force ourselves to think in terms of special relativity when envisaging the behaviour of the elementary particles, especially as all our direct experience is of 'normal' Galilean relativity. What follows is of course only a thumbnail sketch of relativity. Many excellent accounts have been written on the subject, not least of which is that written by Einstein himself.

2.2 Galilean relativity

Galileo's simple example is still one of the clearest descriptions of what relativity is all about. If a man drops a stone from the mast of ship, he will see it fall in a straight line and hit the deck below, having experienced a constant acceleration due to the force of gravity. Another man standing on the shore and watching the ship sail past will see the stone trace out a parabolic path, because, at the moment of release, it is already moving with the horizontal velocity of the ship. Both the sailor and the shoreman can write down their views of the stone's motion using the mathematical equations for a straight line and a parabola respectively. As both sets of equations are describing the same event (the same force acting on the same stone), they are related by transformations between the two observers. These transformations relate the

measurements of position (x^1), time (t^1), and velocity (v^1) in the sailor's coordinate system S^1, with the corresponding measurements (x, t, v) made by the shoreman in his coordinate system S. This situation, assuming that the ship is sailing along the x axis with velocity u, is shown in Figure 2.1.

Important features of the Galilean transformations are that velocity transformations are additive and that time is invariant between the two coordinate frames. Thus if a sailor throws the stone forward at 10 m per second in a ship travelling forward at 10 m per second, the speed of the stone to a stationary observer on shore will be 20 m per second. And if the sailor on a round trip measures the voyage as one hour long, this will be the same duration as observed by the stationary shoreman.

Lest the reader be surprised by the triviality of

Fig. 2.1. The transformations of Galilean relativity.

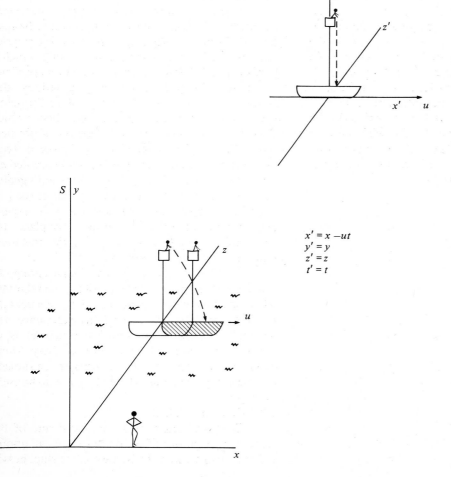

$$x' = x - ut$$
$$y' = y$$
$$z' = z$$
$$t' = t$$

such remarks, let him or her be warned that this is not the case in special relativity. At the high velocities, such as are common in the microworld, velocities do not simply add to give the relative velocity, and time is not an invariant quantity. But before we address these sophistications, let us see how the idea came about.

2.3   The origins of special relativity

The fact that Galilean transformations allow us to relate observations made in different coordinate frames implies that any one inertial frame (a frame at rest or moving at constant velocity) is as good as another for describing the laws of physics. Nineteenth-century physicists were happy that this should apply to mechanical phenomena, but were less happy to allow the same freedom to apply to electromagnetic phenomena, and especially to the propagation of light.

The manifestation of light as a wave phenomenon (as demonstrated in the diffraction and interference experiments of optics) encouraged physicists to believe in the existence of a medium called the ether through which the waves might propagate (believing that any wave was necessarily due to the perturbation of some medium from its equilibrium state). The existence of such an ether would imply a preferred inertial frame, namely, the one at rest relative to the ether. In all other inertial frames moving with constant velocity relative to the ether, measurement and formulation of physical laws (say the force of gravitation) would mix both the effect under study and the effect of motion relative to the ether (say some sort of viscous drag). The laws of physics would appear different in different inertial frames, due to the different effects of the interaction with the ether. Only the preferred frame would reveal the true nature of the physical law.

The existence of the ether and the law of the addition of velocities suggested that it should be possible to detect some variation of the speed of light as emitted by some terrestrial source. As the earth travels through space at 30 km per second in an approximately circular orbit, it is bound to have some relative velocity with respect to the ether. Consequently, if this relative velocity is simply added to that of the light emitted from the source (as in the Galilean transformations), then light emitted

simultaneously in two perpendicular directions should be travelling at different speeds, corresponding to the two relative velocities of the light with respect to the ether, see Figure 2.2.

In one of the most famous experiments in physics, the American physicists, Michelson and Morley set out in 1887 to detect this variation in the velocity of propagation of light. The anticipated variation was well within the sensitivity of their measuring apparatus, but absolutely none was found. This experiment provided clear proof that no such ether exists and that the speed of light is a constant regardless of the motion of the source.

2.4   The Lorentz–Fitzgerald contraction

Around the turn of the century, many physicists were attempting to explain the null result of the Michelson and Morley experiment. The Dutch physicist Lorentz and the Irish physicist Fitzgerald realised that it could be explained by assuming that intervals of length and time, when measured in a

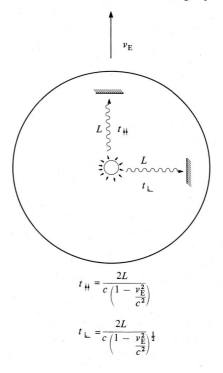

Fig. 2.2 Anticipated variation in the propagation of light reflected to and fro along a distance L due to the earth's motion through space v_E.

$$t_{\parallel} = \frac{2L}{c\left(1 - \frac{v_E^2}{c^2}\right)}$$

$$t_{\perp} = \frac{2L}{c\left(1 - \frac{v_E^2}{c^2}\right)^{\frac{1}{2}}}$$

given frame, appear contracted when compared with the same measurements taken in another frame by a factor dependent on the relative velocity between the two. Their arguments were simply that the anticipated variations in the speed of light were cancelled by compensating changes in the distance and time which the light travelled, thus giving rise to the apparent constancy observed. It is possible to calculate geometrically that an interval of length x measured in one frame is found to be x' when measured in a second frame travelling at velocity v relative to the first where:

$$x = \frac{x'}{\left(1-\frac{v^2}{c^2}\right)^{\frac{1}{2}}}. \tag{2.1}$$

And, similarly, the intervals of time observed in the two frames are related by:

$$t = \frac{t'}{\left(1-\frac{v^2}{c^2}\right)^{\frac{1}{2}}}. \tag{2.2}$$

These empirical relationships, proposed on an *ad hoc* basis by Lorentz and Fitzgerald, suggest that because the 'common-sense' Galilean law of velocity addition fails at speeds at or near that of light, our common-sense perceptions of the behaviour of space and time must also fail in that regime. It was Einstein who, quite independently, raised these conclusions and relationships to the status of a theory.

2.5   The special theory of relativity

The special theory of relativity is founded on Einstein's perception of two fundamental physical truths which he put forward as the basis of his theory:

(1) All inertial frames (i.e. those moving at a constant velocity relative to one another) are equivalent for the observation and formulation of physical laws.

(2) The speed of light in a vacuum is constant.

The first of these is simply the extension of the ideas of Galilean relativity to include the propagation of light, and the denial of the existence of the speculated ether. With our privileged hindsight, the amazing fact of history must be that the nineteenth-century physicist preferred to cling to the idea of

relativity for mechanical phenomena whilst rejecting it in favour of the concept of a preferred frame (the ether) for the propagation of light. Einstein's contribution here was to extend the idea of relativity to include electromagnetic phenomena, given that all attempts to detect the ether had failed.

The second principle is the statement of the far-from-obvious physical reality that the speed of light is truly independent of the motion of the source and so is totally alien to our everyday conceptions. Einstein's achievement here was to embrace this apparently ludicrous result with no qualms. Thus the theory of relativity, which has had such a revolutionary affect on modern thought is, in fact, based on the most conservative assumptions compatible with experimental results.

Given the equivalence of all inertial frames for the formulation of physical laws and this bewildering constancy of the speed of light in all frames, it is understandable intuitively that measurements of space and time must vary between frames to maintain this absolute value for the speed of light. The relationships between measurements of space, time and velocity in different frames are related by mathematical transformations, just as were measurements in Galilean relativity, but the transformations of special relativity also contain the Lorentz–Fitzgerald contraction factors to account for the constancy of the speed of light, see Figure 2.3.

The first feature of the transformations to note is that when the relative velocity between frames is small compared to that of light (i.e. all velocities commonly experienced by humans), then $v/c \approx 0$, and the transformations reduce to the common-sense relations of Galilean relativity.

The unfamiliar effects of special relativity contained in the transformations can be illustrated by a futuristic example of Galileo's mariner: an astronaut in a starship travelling close to the speed of light (c).

Because of the transformations, velocities no longer simply add. If, say, the astronaut fires photon torpedoes forward at speed $1c$ from the starship, which itself may be travelling at $0.95c$, the total velocities of the photon torpedoes as observed by a stationary planetary observer is not the sum, $1.95c$, but is still c, the constant speed of light. Also, time is dilated. So a voyage which to the stationary observer is measured as a given length of time will appear less to the kinetic astronaut.

Another intriguing feature of the transformations is that continued combinations of arbitrary velocities less than c can never be made to exceed c. Thus the transformations imply that continued attempts to add to a particle's velocity (by successive accelerations) can never break the light barrier. Indeed, the transformations themselves do not cater for velocities greater than c, as when $v > c$ the equations become imaginary, indicating a departure from the physical world. Special relativity therefore implies the existence of an ultimate limiting velocity beyond which nothing can be accelerated.

2.6   Mass momentum and energy

If the transformation laws of special relativity show diminishing returns on any attempts to accelerate a particle (by application of some force), it is reasonable to expect some compensating factor to bring returns in some other way, and so maintain energy conservation. This compensating factor is the famous increase in the mass of a particle as it is accelerated to speeds approaching c.

By requiring the laws of conservation of mass

Fig. 2.3. The Lorentz transformations of special relativity.

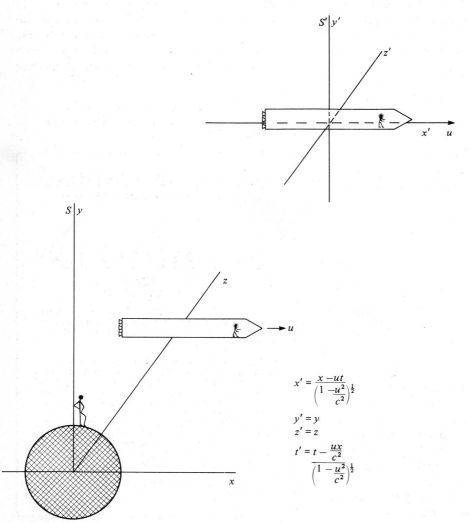

$$x' = \frac{x - ut}{\left(1 - \frac{u^2}{c^2}\right)^{\frac{1}{2}}}$$

$$y' = y$$
$$z' = z$$

$$t' = \frac{t - \frac{ux}{c^2}}{\left(1 - \frac{u^2}{c^2}\right)^{\frac{1}{2}}}$$

and conservation of momentum to be invariant under the Lorentz transformations, it is possible to derive the relationship between the mass of a body m and its speed v,

$$m = \frac{m_0}{\left(1 - \frac{v^2}{c^2}\right)^{\frac{1}{2}}}, \tag{2.3}$$

where m_0 is the mass of the body in a frame in which it is at rest. Multiplying the equation by c^2 and expanding the bracket we obtain:

$$mc^2 = m_0c^2 + \frac{m_0v^2}{2} + \ldots . \tag{2.4}$$

We can identify the second term on the right-hand side of the equation as the classical kinetic energy of the particle. The subsequent terms are the relativistic corrections to the energy whilst the first is describing a quantity of energy arising only from the mass itself.

This is the origin of the mass–energy equivalence of special relativity expressed in the most famous formula of all time:

$$E = mc^2. \tag{2.5}$$

From this formula several others follow immediately. One can be obtained by substituting an expression for the momentum (**p**) into the expansion for m in the above:

$$\mathbf{p} = m\mathbf{v},$$

so

$$E^2 = m_0^2c^4 + p^2c^2. \tag{2.6}$$

For a particle with no rest mass, such as the photon, this gives:

$$\frac{E}{p} = c \tag{2.7}$$

2.7 The physical effects of special relativity

The effects which we have just introduced are all wholly unfamiliar to human experience and this is perhaps one reason why, even today, the reality of special relativity is repeatedly challenged by sceptical disbelievers, see Figure 2.4. But all the effects are real and they can all be measured.

A roll call of the effects of special relativity provides a useful checklist which we should remember when envisaging the behaviour of elementary particles.

2.7.1 *The ultimate speed c*

It is possible to measure directly the velocity of electrons travelling between two electrodes by measuring the time of flight taken. It is observed that the speed does not increase with the energy which the electrons have been given as it would under classical Newtonian theory, but instead tends to a constant value given by c.

2.7.2 *Addition of velocities*

Under special relativity, only when individual velocities are much smaller than c can they be simply added to give the relative velocities. At speeds approaching c, velocities do not add, but combine in a more complicated way so that the total of any combination is always less than c. This can be tested directly by an elementary particle reaction.

Fig. 2.4. Special relativity in trouble? An advertisement from New Scientist magazine 27 May 1982.

DELTA PUBLICATIONS

7305, Aram Nagar, New Delhi-110055, INDIA

A

BIG HOWLER

EINSTEIN'S THEORY OF SPECIAL RELATIVITY

Dr S. P. Gulati & Dr (Mrs) S. Gulati, Associate Professors, Cuttington University College, Liberia; January 1982; 106 pages; Price US\$12·50, STG·6·25 (Air Parcel Postage free).

This book is an open challenging invitation to 'Einsteinians'—particularly so to persons like Professor A. I. Miller of USA who in his recent book 'Albert Einstein's Theory of Special Relativity' (Addison-Wesley, 1981) has undertaken to apotheosize Einstein whose work if not an act of straight plagiarism is definitely 'A BIG HOWLER': infested with infidelities. The 'Transformation Maze' is another interesting feature of the book. Besides, it also contains outlines of the authors' 'SIMILARITY THEORY', perhaps, the only valid alternative.

The book is obtainable either directly from the publisher or through your book-seller or the authors.

One kind of elementary particle we shall encounter is the neutral pion π^0 which often decays into a pair of photons. If the pion is travelling say at $0.99c$ when it emits a photon, we would expect the photon to have a total velocity of $1.99c$ under the laws of Galilean relativity. This is not observed. The photon velocity is measured to be c, showing that very high velocities do not add, but combine according to the formula:

$$v_{\text{total}} = \frac{v_1 + v_2}{\left(1 + \dfrac{v_1 v_2}{c^2}\right)}.$$

2.7.3 *Time dilation*

This is the effect which causes moving clocks to run slow and it has been measured directly in an experiment involving another type of elementary particle. The experiment looks at a species of elementary particle called the mu-meson (or muon), which is produced in the upper atmosphere by the interactions of cosmic rays from outer space. This muon decays into other particles with a distribution of lifetimes around the mean value of 2.2×10^{-6} s when measured at rest in the laboratory. By measuring the number of muons incident on a mountain top, it is possible to predict the number which should penetrate to sea level before decaying. In fact, many times the naive prediction are found at sea level, indicating that the moving particles have experienced less time than if they were stationary. Muons moving at, say, $0.99c$ seem to keep time at only $\frac{1}{4}$ the rate when stationary with respect to us.

2.7.4 *Relativistic mass increase*

The last effect we shall illustrate is the well-known increase in the apparent mass of a body as its velocity increases. This has been measured directly by observing the electric and magnetic deflections of electrons of varying energies, see Figure 2.5.

2.8 Using relativity

As we have seen, relativity tells us how to relate the formulations of physical laws in different frames of reference, but it does not tell us how to formulate them in the first place. This is the rest of physics! In this pursuit, special relativity is introduced by adopting kinematical prescriptions which the dynamical variables must obey.

2.8.1 *Space–time diagrams*

In classical relativity, space and time are entirely separate, but in special relativity they are treated on an identical basis and are mixed together in the Lorentz transformations. So specifying only spatial dimensions in one frame will lead to specification in both space and time dimensions in another. Thus it makes little sense to visualise events as occurring only in space. A better context in which to visualise them is space–time. Space–time diagrams can be used to display events at the expense, for the purposes of visualisation, of making do with only one, or possibly two, spatial dimensions, see Figure 2.6.

Fig. 2.5. Relativistic mass increase as a function of velocity.

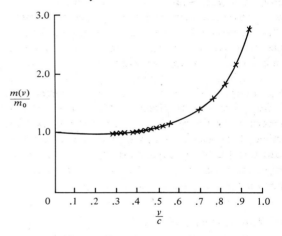

Fig. 2.6. A space–time diagram particle collision (*b*) shown sequentially in (*a*).

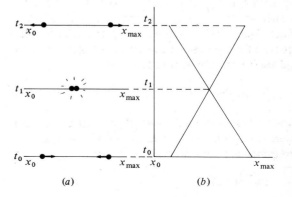

2.8.2 *Four-vectors*

Just as ordinary vectors (**x**) (three-vectors) define the components of a position of velocity in ordinary space, we can define four-vectors (**x**, ct) to define an event in space–time. The fourth co-ordinate is the time coordinate multiplied by c to give an equivalent distance, so matching the other three distance components.

The benefit of writing equations in three-vector form is to ensure their covariance under spatial rotations. (Covariance is not quite the same as invariance which means that absolutely nothing changes. Covariance means that both sides of an equation change in the same way, preserving the validity of the equation.) This permits freedom in the orientation of the coordinate system employed and also ensures the conservation of angular momentum (see Chapter 11). If we can write the laws of physics in four-vector form, then the benefit is that the laws will be covariant under rotations in space–time (which are equivalent to the Lorentz transformations of special relativity).

In addition to the position-three vector (**x**), the momentum of a particle is also a vector quantity (**p**). By examining the effects of the Lorentz transformations on the momentum and the energy of a particle, it is possible to form a four-vector from the quantities (**p**, E/c). This four-vector is used to specify the dynamic state of a particle. It does not specify an event in space–time.

2.8.3 *Relativistic invariants*

Although special relativity illustrates how perceptions of space and time may vary according to the observer's frame, it also accommodates absolutely invariant quantities, which we might expect to vary under Galilean relativity. The speed of light in a vacuum is the obvious invariant upon which the theory is founded. Another quantity is the square of the space–time interval between an event and the origin of the coordinate system,

$$s^2 = (ct)^2 - x^2 = (ct')^2 - x'^2. \tag{2.8}$$

This is just a special case of the square of the space–time interval between two events, which is the difference between their four-vectors,

$$\Delta s^2 = (c\Delta t)^2 - \Delta x^2.$$

Another invariant quantity is the rest mass of a given material particle. All observers will agree on the mass of the same particle at rest in their respective frames.

$$m_0^2 = \frac{E^2}{c^4} - \frac{p^2}{c^2}. \tag{2.9}$$

Relativistic invariants are useful in high-energy physics because, once measured, their value will be known in all other circumstances. Here it is worth appreciating that high-energy experiments regularly exercise the idea of Lorentz transformations. One experiment may arrange for two protons travelling with equal energies in opposite directions to collide head-on, whilst another experiment may collide moving protons and a stationary target. The centres of mass of the two experiments will be moving relative to one another with some velocity which is likely to be an appreciable fraction of the speed of light. This will require the Lorentz transformations to relate measurements in the two experiments.

This concludes our brief sketch of special relativity and now we pass on to the second of the two great pillars of twentieth-century physics: quantum mechanics.

3

Quantum mechanics

3.1 Introduction

It is fascinating to reflect on the fact that both quantum mechanics and special relativity were conjured into being in the first five years of this century, and interesting to compare the development of the two. Whereas special relativity sprang as a complete theory (1905) from Einstein's genius, quantum mechanics emerged in a series of steps over a quarter of a century (1900–25). One explanation of this is that whereas in special relativity the behaviour of space and time follow uniquely from the two principles, in quantum theory there are no such principles which, known at the beginning, allow the derivation of all the subsequent consequences. Rather, each of the steps is a fresh hypothesis based on, or predicting, some new experimental facts and these do not necessarily follow logically one from another, still less from just one or two fundamental principles. So quantum mechanics emerged, hypothesis hand-in-hand with experiment, over 25 years or so. As indicated by the sub-headings of this chapter, most of the steps in the progression can be associated closely with just one man, and we will use the examination of each of these in turn as our introduction to quantum mechanics.

3.2 Planck's hypothesis

As we have mentioned in Section 1.5.2, quantum theory came into being when Max Planck attempted to explain the interaction of light with matter. That is, for instance, how hot metal emits light and how light is absorbed by matter.

Using the well-known and highly trusted classical theories of thermodynamics and electromagnetism, Planck derived a formula describing the power emitted by a body, in the form of radiation, when the body is heated. To find the total power radiated, it is necessary to integrate over all the possible frequencies of the emitted radiation. But when Planck tried to do this using his classical formula, he found that the total radiated power was predicted to be infinite – an obviously nonsensical prediction!

Planck was able to avoid this conclusion only by introducing the concept of a minimum amount of energy which can exist for any one frequency of the radiation, a quantum. By assuming that light can be emitted or absorbed by matter only in multiples of the quantum, Planck derived a formula which gives the correct prediction for the total amount of power emitted by a hot body. A convenient analogy here may be the economic wealth of an individual, which is normally thought of as a continuously variable quantity. Yet when the individual is in economic interaction (i.e. he goes shopping), his or her wealth is quantised in multiples of the smallest denomination coin available. The minimum quantum of energy E, allowed at a given frequency v, is given by Planck's formula

$$E = hv, \qquad (3.1)$$

where h is Planck's quantum constant with dimensions of energy per frequency and the minute value of 6.625×10^{-34} joule seconds. The appearance of Planck's constant in the equations of physics is a valuable diagnostic device. When we set $h = 0$, then we are ignoring the existence of the quantum and so should recover the results of classical physics. However, when we examine formulae (or parts of formulae) which are proportional to h, then we are looking at wholly quantum effects which would not be predicted by classical physics.

3.3 Einstein's explanation of the photoelectric effect

The next major step in quantum theory was taken by Einstein in the same year as his formulation of special relativity. This was his explanation of the photoelectric effect, or how metal can be made to emit electrons by shining a light on it. Planck had suggested that only light in interaction with matter would reveal its quantum behaviour at low energies. Again it was left to Einstein to generalise the idea (as he had generalised the ideal of relativity to include electromagnetism). He proposed that all light exists in quanta and set out to show how this might explain the photoelectric effect.

He assumed that the electrons need a definite amount of energy to escape from the metal. If the light of a given colour which is shone on the metal consists of a large number of quanta, each of energy hv, then quanta which collide with the electrons provide them with the energy they need to escape. The electrons will pop out of the metal with an energy which is the difference between that of the quanta and that needed to escape the surface. If the light is below a certain frequency, then no matter how much of it is used, no single quantum will be able to give an electron enough energy to escape. Ignoring multiple quanta–electron collisions, no electrons will emerge. But if the frequency of the light is increased, scanning up the spectrum from red to blue, the electrons will suddenly appear when the quanta have just enough energy to liberate them. As the frequency is increased further still, the electrons will be ejected with higher and higher energies.

This picture exactly fits the experimental facts of the photoelectric effect discovered in 1902 by Lenard. These are that the energy of the electrons emitted depends only on the frequency of the light and not on the intensity (the number of quanta), and that the number of electrons emitted depends only on the intensity but not the frequency.

Einstein's explanation of the photoelectric effect confirmed the quantum theory of light (and won him the Nobel prize). This resurrection of a corpuscular theory of light causes immediate conceptual problems because light is quite demonstrably also a continuous wave phenomenon (as demonstrated by diffraction and other interference experiments). It appears to be both a discrete particle (a *photon*) and an extended wave! How can this be?

Resolution of this apparent paradox requires the introduction of a new entity which reduces to both particle and wave in different circumstances. This entity turns out to be a *field* which we shall

discuss further in Chapter 4. But before going on to this we will come to appreciate that not only light is subject to such schizophrenic behaviour.

3.4 Bohr's atom

We saw in Chapter 1 how Rutherford's scattering experiments led to a picture of the atom in which the light, negatively charged electrons orbit the small, massive, positively charged nucleus located in the centre, the vast majority of the volume of the atom being empty space. This appealing picture has fundamental difficulties. Firstly, in the classical theory of electrodynamics, all electric charges which experience an acceleration should emit electromagnetic radiation. Any body constrained to an orbit is subject to an acceleration by the force which gives rise to the orbit in the first place. Thus the electrons in Rutherford's atom should be emitting radiation constantly. This represents a loss of energy from the electrons which, as a result, should spiral down into lower orbits and eventually into the nucleus itself. This 'radiation collapse' of atoms is an inescapable consequence of classical physics and represents the failure of the theory in the atomic domain. Another problem of the Rutherford atom is to explain why all the atoms of any one element are identical. In classical physics, no particular configuration of electronic orbits is predicted other than on the grounds of minimising the total energy of the system. This does not explain the identity of the atoms of any element. A fundamentally new approach is needed to describe the Rutherford atom.

It was the Danish physicist Niels Bohr who in 1913 suggested a new quantum theory of the atom, which, at a stroke, dismissed the problem of the radiation collapse of atoms, explained the way in which light is emitted from atoms and incorporated the new quantum ideas of Planck and Einstein.

Bohr's basic hypothesis was the simplest possible application of the quantum idea to the atom. Just as Planck had hypothesised that light exists only in discrete quanta, so Bohr proposed that atoms can exist only in discrete quantum states, separated from each other by finite energy differences, and that *when in these quantum states the atoms do not radiate*. A simple way to think of these quantum states is as a set of allowed orbits for the electron around the nucleus, the space between the orbits being forbidden to the electrons.

The allowed orbits are specified as those in which the orbital angular momentum of the electron is quantised in integral units of Planck's quantum constant divided by 2π and denoted $\hbar$. It may seem odd that angular momentum should be one of the few quantities to be quantised (like energy and electric charge but unlike mass, linear momentum and time). But we may have suspected as much on first meeting Planck's constant. Its rather unusual units of energy per frequency are in fact identical to the dimensions of angular momentum.

Although the atom is assumed not to radiate light when all its electrons are safely tucked into their quantum orbits, it will do so when an electron makes a transition from one of the allowed orbits to another. This process of emission should explain the behaviour of light observed in the real world. The light from a gas discharge lamp, say a neon or mercury vapour tube has a distinctive appearance. The atoms in a gas or vapour are widely separated and interact with each other relatively seldom. This means that the light they emit will be characteristic of the particular atoms involved. It is a mixture of just a few separate frequencies which can be split up by a prism. The resulting spectrum of frequency lines is a unique property of the element which is emitting the light, see Figure 3.1. Late in the

nineteenth century, physicists such as Balmer, Lyman and Paschen looked at the spectra of many different elements and noted that they all fall into simple mathematical patterns – with several discrete patterns per element. These patterns had long defied explanation, essentially because they defy the smooth way in which quantities vary in classical physics. But with the quantum theory, Bohr was able to forward a convincing explanation of the origin of these lines. Each pattern of frequency lines represents the energy difference between a particular quantum state and all the others in the atom from which the electron can reach that state by emitting light, see Figure 3.2.

With Bohr's model of the atom, physicists were able to calculate, in great detail, many of the spectroscopic results obtained by the experimenters of previous decades. On the basis of this understanding of atoms, Bohr himself was able to propose a tentative explanation of Mendeleeff's periodic table of elements. The periodic table, which classifies the elements into groups reflecting their chemical behaviour, is explained by the way the electronic orbits are filled in the different elements. The chemical properties of an element are determined predominantly by the number of electrons in its outermost orbit, and so by proposing the electronic orbital structures of the elements it is possible to reproduce the pattern of the table, see Figure 3.3.

Whilst the Bohr atom was an enormous step forward, and the concept of electronic orbits is a mental crutch for our imaginations operating so far

Fig. 3.1. The characteristic spectrum of a gas discharge lamp.

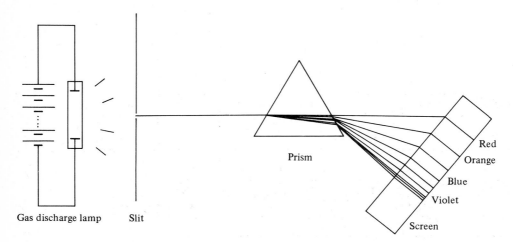

Gas discharge lamp   Slit

Prism

Red
Orange
Blue
Violet

Screen

Fig. 3.2. The discrete patterns of frequency lines in a given element (such as hydrogen) arise from transitions into each available state from all the others. Each state is labelled by its Bohr orbital quantum number *n*.

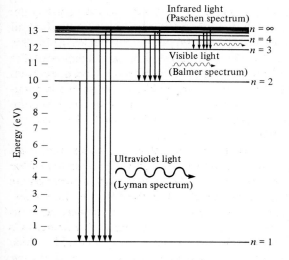

beyond their normal domain, it is important to realise that it is only the simplest quantum model of the atom and that more-sophisticated portrayals of electronic behaviour are necessary, as we are about to see.

3.5   de Broglie's electron waves

The next major conceptual advance in quantum theory came much later, in 1924. The young French physicist Louis de Broglie suggested in his doctoral thesis that just as light waves could act like particles in certain circumstances, so too could particles manifest a wavelike behaviour. In particular, he proposed that the electrons, which had previously been regarded as hard, impenetrable, charged spheres could in fact behave like extended waves undergoing diffraction and interference phenomena just like light or water waves.

According to de Broglie, the wavelength of a particle wave is inversely proportional to its momentum, the constant of proportionality being Planck's quantum constant.

Fig. 3.3. A fragment of the periodic table and the associated electronic orbital structure.

$$\lambda = \frac{h}{p}. \tag{3.2}$$

Chemical group number

I	II	III	IV	V	VI	VII	VIII

Hydrogen							Helium
H							He

Lithium	Beryllium	Boron	Carbon	Nitrogen	Oxygen	Flourine	Neon
Li	Be	B	C	N	O	F	Ne

So the higher the momentum of a particle, the smaller its wavelength. It is worth appreciating that de Broglie's hypothesis applies to all particles, not just to electrons and the other elementary particles. For instance, a billiard ball rolling across the table top will have a wavelength, but because Planck's constant is so minute and the ball's momentum is so comparatively large, the billiard ball wavelength is about 10^{-34} m. This, of course, is many orders of magnitude different from the typical dimensions of billiards, and so the wave character of the ball never reveals itself. But for electrons, their typical momenta can give rise to wavelengths of 10^{-10} m which are typical of atomic distance scales. So electrons may be expected to exhibit a wavelike character during interaction with atomic structures. This was first shown in 1927 by the American physicists Clinton Davisson and Lester Germer who demonstrated that electrons can be diffracted through the lattice structure of a crystal in a fashion similar to the diffraction of light through a grating.

de Broglie's hypothesis also provided the first rationale for Bohr's model of the atom. The existence of only certain specific electronic orbits can be explained by allowing only those orbits which contain an integral number of de Broglie wavelengths. This reflects the momentum of the electron involved (and so the energy of the orbit), see Figure 3.4.

Adoption of de Broglie's idea requires the comprehensive assimilation of particle–wave duality. For any entity in the microworld, there will be situations in which it is best thought of as a wave and situations in which it is best thought of as a particle. Neither is a truer representation of reality than the other, as both are the coarse product of our human macroscopic imaginings.

The advent of de Broglie's ideas marks the spark which started the intellectual bush fire of quantum theory proper. Up to the early 1920s, quantum theory was a series of prescriptions (albeit revolutionary ones) but not a dynamical theory of mechanics to transcend that of Newton. The second wave of the quantum revolution (1924–27) was to provide just such a theory.

3.6　Schrödinger's wavefunction

Following on directly from de Broglie's ideas, the Austrian physicist, Erwin Schrödinger developed the idea of particle waves into a wave mechanics proper. Schrödinger's starting point was essentially the wave equation describing the behaviour of light waves in space and time. Just as this is the accurate representation of optical phenomena (which can be described approximately by the light *rays* of geometrical optics), Schrödinger formulated a matter wave equation which he put forward as the accurate representation of the behaviour of matter (which is described approximately by the particle dynamics). Schrödinger's equation describes a particle by its wavefunction, denoted ψ, and goes on to show how the particle wavefunction evolves in space and time under a specific set of circumstances.

One such circumstance of very great interest is that of a single electron moving in the electric field of a proton. Using his wave equation, Schrödinger was able to show that the electron wavefunction can assume only certain discrete energy levels, and that those energy levels are precisely the same as the energies of the electronic orbits of the hydrogen atom, postulated earlier by Bohr.

The particle wavefunction is an extremely significant concept which we will use frequently in the coming chapters. It is a mathematical expression describing all the observable features of a particle. Collisions between particles are no longer necessarily viewed as some variant of billiard-ball behaviour but, instead, as the interference of wavefunctions giving rise to effects rather like interference phenomena in optics.

But now that we have introduced the particle

Fig. 3.4. Allowed orbits are explained as containing an integral number of de Broglie wavelengths.

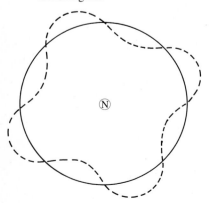

wavefunction, and claimed that an equation governing it can predict the behaviour of particles, what exactly is its significance? Should we think of an electron as a localised ball of stuff, or as some extended wave? And if a wave, what is doing the waving? After all, there is no such thing as a light wave, it is a handy paraphrase for time- and space-varying electric and magnetic fields. What then, is a matter wave?

Before we answer these intuitive questions, we need one more principle of quantum theory. This is the uncertainty principle which the German physicist Werner Heisenberg derived from his alternative formulation of a quantum mechanics, developed simultaneously with Schrödinger's wave formulation, but from a very different starting point.

3.7 Heisenberg's mechanics and the uncertainty principle

Heisenberg took as his starting point the quantum state of the system under consideration (e.g. a single electron, an atom, a molecule etc.), and argued that the only sensible way to formulate a mechanics of the system was by modelling the act of observation on it. Here, by the word 'observation' we mean any interaction experienced by the system, such as the scattering off it of light or of an electron. In the absence of any interaction, the system would be totally isolated from the outside world and so totally irrelevent. Only by some form of interaction or observation does the system exist (to all practical intents and purposes).

Heisenberg's approach is the literal manifestation of Wittgenstein's parting philosophical rejoinder, 'concerning that of which we cannot speak, we must pass over in silence'. We can speak (or write equations) only of what we observe, and so observation is to have pride of place in quantum theory.

Heisenberg represented observations on a system as mathematical operations on its quantum state. This allowed him to write equations governing the behaviour of a quantum system and so led to results which were identical to the somewhat more accessible wave mechanics of Schrödinger (say in predicting the energy levels of the hydrogen atom). The equivalence of the two approaches can be appreciated by realising that the expressions Heisenberg used to represent the observations are

differential operators and that they act on the quantum state, which is represented by the wavefunction of the system. So this approach will result in a differential equation in the wavefunction ψ, identical to the wave equation which Schrödinger obtained by analogy with the wavefunction for light.

Heisenberg's uncertainty principle results from the realisation that any act of observation on the quantum system will disturb it, thus denying perfect knowledge of the system to the observer. This is best illustrated by analysis of what would happen if we were to attempt to observe the position of an electron in an atomic orbit by scattering a photon off it, Figure 3.5. The photon's wavelength is related to its momentum by the same equation as for any other particle:

$$\lambda = \frac{h}{p}.$$

Fig. 3.5. A long-wavelength (low-momentum) photon can give only a rough estimate of the position of the electron but does not disturb the atom very much. A short-wavelength (high-momentum) photon localises the electron more accurately, but causes great disturbance.

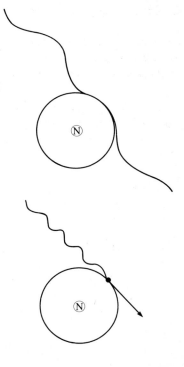

So the greater the photon's momentum, the shorter its wavelength and vice versa. If then we wish to determine the position of the electron as accurately as possible, we should use the highest possible momentum (shortest wavelength) photon, as it is not possible to resolve distances shorter than the wavelength of the light used. However, by using a high-momentum photon, although we will gain a good estimate of the electron's position at the instant of measurement, the electron will have been violently disturbed by the high momentum of the photon and so its momentum will be very uncertain. This is the essence of Heisenberg's uncertainty principle. Knowledge of any one parameter implies uncertainty of some other so-called 'conjugate' parameter. This is expressed mathematically by requiring that the product of the uncertainties in the two conjugate parameters must always be greater than or equal to some small measure of the effect of measurement. Not surprisingly, this measure turns out to be none other than Planck's ubiquitous constant,

$$\Delta p \Delta x \geqslant \hbar, \text{ with } \hbar = \frac{h}{2\pi}.$$

A similar trade-off occurs when attempting to measure the energy of a quantum system at a given time. An instantaneous measurement implies a high-frequency probe (one wavelength over in a short time), but this means a high-energy probe which will mask the energy of the quantum state itself. Conversely, a very low-energy probe which will not unduly affect the energy of the quantum state, implies a low frequency probe, which means the time to be associated with the measurement is uncertain, thus:

$$\Delta E \Delta t \geqslant \hbar.$$

Heisenberg's uncertainty principle is an enormously powerful result when we realise that the uncertainty in a quantity provides a good guide to its minimum value. For instance, if we know that the uncertainty in a particle's lifetime is 1 s, then the lifetime is unlikely to be less than $\frac{1}{2}$ s as the uncertainty could not otherwise be accommodated. Similarly, if we know that a particle is confined to a small volume (say the nucleus $\Delta x \approx 10^{-15}$ m), then we can conclude that the momentum of the particle must be greater than

$$p_{\min} \approx \frac{\Delta p}{2} \approx \frac{\hbar}{2\Delta x} \approx 100 \text{ MeV}/c.$$

If the particle is confined to the nucleus, then this is a reasonable guide to the strength (energy) of the force which is keeping it there.

Armed with these ideas, we can turn to the thorny problem of just what is a matter wave.

3.8 The interpretation of the wavefunction ψ

Firstly, we can address the question of whether an electron is to be regarded as a localised ball or an extended wave. Which of these two descriptions applies is very much a matter of the circumstances the electron finds itself in, see Figure 3.6.

For an electron which is travelling through space with a definite momentum ($\Delta p = 0$) and so isolated from all interactions, the uncertainty in its position is infinite. Thus its wavefunction is a sine wave of definite wavelength extending throughout space. The electron is in no sense a localised particle. If an electron is vaguely localised, say we know it has disturbed an atom, then with Δx as the dimension of the atom, we know that there will be an uncertainty Δp in the electron momentum (due to its interaction with the atom) and so a spread in the

Fig. 3.6. A particle's wavefunction reflects its localisation (see text).

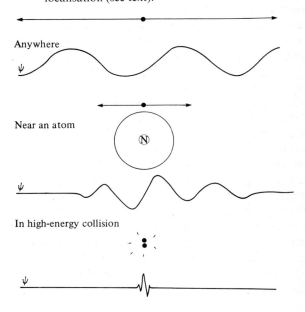

wavelength of the wavefunction, $\Delta\lambda = h/(\Delta p)$. This spread in wavelengths (frequencies) causes the formation of a localised wavepacket in the wavefunction reflecting the rough localisation of the electron.

When the electron is very specifically localised, say in a quasi-point-like, high-energy collision with another particle, then the uncertainty in its momentum (and so the spread in the wavelength components of the wavefunction) is large, and the wave-packet becomes very localised, in which case it is sensible to regard the electron as a particle.

This picture of the electron wave makes rather a nonsense of the simple Bohr picture of the orbiting electrons. The dimensions of the electronic wavefunction are comparable to that of the atom itself. Until some act of measurement localises the electron more closely, there is no meaning to ascribing any more detailed a position for the electron. However, this explanation is not altogether satisfactory as it stands, as we have left the electron with a rather poorly defined role in the atom. Progress in understanding this aspect is related to our other outstanding question about the wavefunction; what is it?

In 1926 the German physicist Max Born ventured the suggestion that the square of the amplitude of the wavefunction at any point is related to the probability of finding the particle at that point. The wavefunction itself is proposed to have no direct physical interpretation other than that of a 'probability wave'. When squared, it gives the chance of finding the particle at a particular point on the act of measurement.

So the location of the electron in the atom is not wholly indeterminate. The solution to Schrödinger's equation for an electron in the electrical field of the proton will give an amplitude for the wavefunction as a function of distance from the proton (as well as the energy levels mentioned earlier). When squared, the amplitude gives the probability of finding the electron at any particular point. Thus we can give only a probability for finding an electron in its Bohr orbit, a probability for determining its position within the orbit and probabilities for finding it in the space between orbits. There is even a small probability of this so-called orbital electron existing actually inside the nucleus!

3.9   Electron spin

Having just developed a rather sophisticated picture of the electronic wavefunction, we shall immediately retreat to the comfortingly familiar picture of Bohr's orbital atom to explain the next important development in quantum theory!

By 1925, physicists attempting to explain the nature of atomic spectra had realised that not all was correct. Where, according to Bohr's model, just one spectral line should have existed, two were sometimes found very close together. To explain this and other similar puzzles, the Dutch physicists Sam Goudsmit and George Uhlenbeck proposed that the electron spins on its axis as it orbits around the nucleus (just as the earth spins around the north–south axis as it orbits around the sun), see Figure 3.7.

The splitting of the spectral lines is explained by the existence of magnetic effects inside the atom. The electron orbit around the nucleus forms a small loop of electric current and so sets up a magnetic field; the atom behaves like a small magnet. The spin of the electron is an even smaller loop of electric current and this sets up another magnetic field which is referred to as the 'magnetic moment of the electron'. This can either add to, or subtract from, the main magnetic field of the atom, depending on the way in which the electron is spinning. This will lead to a slight difference in the energy of the

Fig. 3.7. In the orbital picture of a particle electron, the electron spins on its own axis.

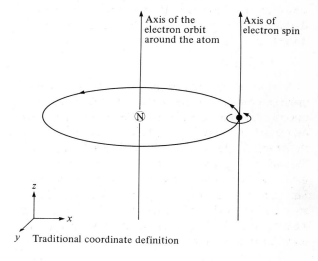

y   Traditional coordinate definition

electronic orbit for the different spins of the electron, and will result in the splitting of the spectral line associated with the Bohr orbit.

The above is a nice classical picture, but it has its limitations. The fact that the spectral line splits into just two components indicates that the electron cannot be spinning around at any arbitrary angular momentum but must be such that it has just two values along the line of the atom's magnetic field (or, in the case of a free electron, along the line of any applied magnetic field). The components of the spin in this direction are referred to as the 'z components' or the 'third components' of spin and are measured to be quantised in half-integral units of Planck's quantum constant (divided by 2π),

$$s_z = \pm \frac{\hbar}{2}.$$

Although the picture of the electron as a spinning ball is attractive, it is important to remember that it is simply a model. In fact, electron spin is purely a quantum concept (it is directly proportional to $\hbar$). We must be prepared to think also of the electron as an extended wave which carries a quantum of intrinsic angular momentum, just like its quantum of electric charge.

Other particles also carry spin. The proton and the neutron carry spin quanta which are half-integral multiples of Planck's constant, just like the electron. The photon also has spin, but the quantum is of one complete unit $\hbar$. As the photon is simply a packet of electric and magnetic fields this shows that intrinsic angular momentum can be a feature of purely non-material fields. As we shall soon see, the difference in particle spins is very important. On a fundamental level it gives a method of categorising the behaviour of the wavefunctions of particles under the Lorentz transformations of special relativity (a connection we will discuss further in Chapter 11). On a practical level, it signifies very different behaviours of ensembles of particles (see next section).

3.10 The Pauli exclusion principle

A straightforward look at the Bohr model of the atom tells us that some fundamental principle must be missing. For there is seemingly nothing to prevent all the electrons of any atom from performing the same orbit. Yet we know that a typical atom will have its electrons spread over several different orbits. Otherwise, transitions between them would be rare, in contradiction to the observations of atomic spectra. So some rule must keep the electrons spread out across the orbits of the atom.

In 1925 the Austrian physicist Wolfgang Pauli derived the principle that no two electrons can simultaneously occupy precisely the same quantum state (i.e. have identical values of momentum, charge and spin in the same region of space). He reached this conclusion after examining carefully the atomic spectra of helium. He found that transitions to certain states were always missing, implying that the quantum states themselves were forbidden. For instance, the lowest orbit (or ground state) of helium in which the two electrons have the same value of spin is not present. But the state in which the two electron spins are opposite is observed.

The power of this principle in atomic physics can hardly be overstated. Because no two electrons can exist in the same state, the addition of extra orbital electrons will successively fill up the outer-lying electron orbits and will avoid overcrowding in the lowest one. Just two electrons are allowed in the ground state because the only difference can be the two values of spin available. More electrons are allowed in the higher orbits because their quantum states can differ by a wider range of orbital angular momenta around the nucleus (which also turns out to be quantised). It is the Pauli exclusion principle which is responsible for the chemical identities of all atoms of the same element, as it is this principle which determines the allowed arrangements of the atomic electrons.

Although we have focussed on the atom, the exclusion principle applies to any quantum system, the extent of which is defined principally by the wavefunctions of the component particles. In the case of totally isolated electrons of definite momentum whose wavefunctions extend over all space, the exclusion principle means that only two electrons with opposite spins can have the same momentum. In the case of electrons confined to a crystal (i.e. electrons whose wavefunctions extend over the dimensions of the crystal), the rule will apply to all electrons in the crystal.

Pauli's exclusion principle can be expressed alternatively in terms of the behaviour of the wavefunction of a quantum system. Although we have talked so far only of the wavefunctions of individual particles, these can be aggregated for any quantum system to give a wavefunction describing the whole system. For example, the total wavefunction of the helium atom can describe the behaviour of two electrons at the same time. Just as the wavefunction of a single electron is a wavepacket reflecting the localisation of the electron, a double-electron wavefunction will contain two wave-packet humps reflecting the localisations of the two electrons. The consequence of the exclusion principle for any multiple-electron wavefunction is that it must change sign under the interchange of any two electrons. Wherever the amplitude is positive it must become negative and vice versa. The wave-function is said to be antisymmetric under the interchange of two electrons. This effect can be understood by considering the two-electron helium atom. If the two electrons cannot be in the same quantum state (differing spins), they must be distinguishable under interchange and the wavefunction must signal this. On the other hand, all we are doing is relabelling the electrons and this should make no difference to the physical results (e.g. energy levels and probability densities). The antisymmetry of the wavefunction allows just this. As all physical quantities are proportional to its square, changing only its sign will make no difference.

Particles such as the electron and the proton with spin $\frac{1}{2}\hbar$ (and other more exotic particles that we shall meet with other half-integral spins $\frac{3}{2}\hbar$, $\frac{5}{2}\hbar$, . . .) obey the exclusion principle, give rise to antisymmetric wavefunction under the interchange of two identical such particles and are referred to as *fermions*. This is because an ensemble of fermions obey statistics governing their dynamics, which were first formulated by the Italian physicist, Enrico Fermi, and the Englishman, Paul Dirac. Fermi–Dirac statistics show how momentum is distributed amongst the particles of the ensemble. Because of the exclusion principle in any quantum system, there is a limit to the number of particles which can adopt any particular value of momentum and so this leads to a wide range of momentum carried by the particles. Particles such as the photon with spin $\hbar$ (and other particles we shall meet with integral spins, 0, $\hbar$, $2\hbar$, $3\hbar$, . . .) do not obey the exclusion principle and are call *bosons*. Their wavefunction

does not alter under the interchange of two particles. An assembly of bosons obeys dynamical statistics first formulated by the Indian physicist Satiendranath Bose and Albert Einstein. In Bose–Einstein statistics there is no limit to the number of particles which can have the same value of momentum, and this allows the assembly of bosons to act coherently, as in the case of laser light.

This last principle concludes our whistle-stop tour of quantum mechanics. Although brief, the tour has included most of the new concepts introduced by the theory. For the purposes of the rest of the book, the most important of these is the wavefunction interpretation of a particle, although we will use the uncertainty and exclusion principles from time to time. As in the case of relativity, it is a constant challenge to shrug off our everyday imaginings in the microworld and learn to think in terms of these unfamiliar ideas. But before we are quite ready to approach the subject we must look at what happens when relativity and quantum mechanics are put together.

4

Relativistic quantum theory

4.1 Introduction

Quantum mechanics, just like ordinary mechanics and electrodynamics, must be made to obey the principles of special relativity. Because the entities (particles, atoms, etc.) described by quantum theory quite often travel at speeds at or near c, then this becomes an essential requirement. Special relativity will not just give corrections to conventional Newtonian mechanics, but will dictate dominant, unconventional relativistic effects.

We will see that the synthesis of relativity with quantum theory predicts wholly new and unfamiliar physical consequences (antimatter). This requires us to develop a new way of looking at matter (quantum fields). If we can then go on to develop the mechanics of interacting quantum fields, this will provide us with the most satisfactory description of the behaviour of matter (both the conventional matter we have discussed so far, and the unconventional antimatter we have had to introduce along the way).

4.2 The Dirac equation

At the same time as Schrödinger and Heisenberg were formulating their respective versions of the quantum theory, Paul Dirac was attempting the same task. But, in addition, he was concerned that the quantum theory should manifestly respect Einstein's special relativity. This implies two distinct requirements: firstly, that the theory must predict the correct energy–momentum relation for relativistic particles,

$$E^2 = m_0^2 c^4 + p^2 c^2$$

and, secondly, that the theory must incorporate the phenomenon of electron spin in a Lorentz covariant fashion.

In one of the most celebrated brainstorming sessions of theoretical physics, Dirac simply wrote down the correct equation! He was guided in this task by realising that Schrödinger's equation for the electronic wavefunction cannot possibly satisfy the requirements of special relativity because time and space enter the equation in different ways (as first- and second-order derivatives respectively). Schrödinger's equation is perfectly adequate for particles moving with velocities much less than c, and it predicts the correct Newtonian energy–momentum relationship for particles,

$$E = \frac{p^2}{2m} = \frac{mv^2}{2}.$$

But because space and time are not treated correctly, it does not predict the correct relativistic relationships or incorporate energy–mass equivalence.

In the spirit of special relativity, Dirac sought an equation treating space and time on an equal basis. In this he succeeded, but found that in doing so the electron wavefunction ψ could no longer be a simple number. Incorporating time and space on an equal basis requires the electron wavefunction ψ to contain two separate numbers which may be interpreted as reflecting the probabilities that the electron is spin up (with spin quantum $\hbar/2$) or spin down (with spin quantum $-\hbar/2$). Thus ψ is written as a two-component '*spinor*' $\psi = \begin{pmatrix} a \\ b \end{pmatrix}$. In fact, in the full theory it is a four-component object, for reasons which will become clear in the next section.

In attempting to incorporate special relativity into quantum mechanics it is necessary to invent electron spin! It is fascinating to wonder whether if electron spin had not been proposed and discovered experimentally, it would have been proposed theoretically on this basis. And if so, if it would still have become known as electron spin or if it would have been introduced as some sort of 'Lorentz charge'.

Dirac's equation can be used for exactly the same purposes as Schrödinger's, but with much greater effect. In Section 3.9 we saw that the spin of the electron gives rise to a splitting in the energy levels of the hydrogen atom. This is because the magnetic moment of the electron may either add to, or subtract from, the magnetic field set up by the electron's orbital angular momentum. It was noticed in experiments that the half-integral unit of spin angular momentum $\hbar/2$ produced as big a magnetic moment as a whole integral unit of orbital angular momentum (i.e. spin is twice as effective in producing a magnetic moment as is orbital angular momentum). This is quantified by ascribing the value of 2 to the gyromagnetic ratio (the g factor) of the electron. This is effectively the constant of proportionality between the electron spin and the magnetic moment resulting. In non-relativistic quantum mechanics, $g = 2$ is an empirical fact. With the Dirac equation, it is an exact prediction.

The Dirac equation can also explain the fine splitting and hyperfine splitting of energy levels within the hydrogen atom. These result from the magnetic interactions between the electron's orbital angular momentum, the electron spin and the proton spin.

4.3 Antiparticles

One immediate consequence of predicting the relativistic relationship between energy and momentum for the electron wavefunction is that the Dirac equation seems to allow the existence of both positive- and negative-energy particles:

$$E = \pm (m_0^2 c^4 + p^2 c^2)^{\frac{1}{2}}.$$

In an amazing feat of intellectual bravado, Dirac suggested that this prediction of negative-energy particles was not rubbish but, instead, the first glimpse of a hidden universe of antimatter.

The concept of negative-energy entities is wholly alien to our knowledge of the universe. All things of physical significance are associated with varying amounts of positive energy. So Dirac did not ascribe a straightforward physical existence to these negative-energy electrons. Instead, he proposed an energy spectrum containing all electrons in the universe, see Figure 4.1. This spectrum consists of all positive-energy electrons which inhabit a band

of energies stretching from $m_0 c^2$, the rest mass, up to arbitrarily high energies. These are the normal electrons which we observe in the laboratory and whose distribution over the energy spectrum is determined by the Pauli exclusion principle. Dirac then went on to suggest that the spectrum also contains the negative-energy electrons which span the spectrum from $-m_0 c^2$ down to arbitrarily large negative energies. He proposed that these negative-energy electrons are unobservable in the real world. To prevent the real, positive-energy electrons simply collapsing down into negative-energy states, it is necessary to assume that the entire negative-energy spectrum is full and that double occupancy of any energy state in the continuum is prevented by the Pauli exclusion principle. No electrons inhabit the energy gap between $-m_0 c^2$ and $m_0 c^2$. (There are no fractions of electrons in evidence.)

Viewed picturesquely, it is as if the world of physical reality conducts itself whilst hovering over an unseen sea of negative-energy electrons.

But if this sea of negative-energy electrons is to remain unseen, what is its effects on the everyday world? The answer to this is that elementary particle interactions of various sorts can occasionally transfer enough energy to a negative-energy electron to boost it across the energy gap into the real world. For instance, a photon with energy $E \geqslant 2m_0 c^2$ may collide with the negative-energy electron and so promote it to reality. But this cannot be the end of the story, as we seem to have created a unit of electrical charge, whereas we are convinced that this is a quantity which is conserved absolutely. Also, we started out with a photon of energy $E \geqslant 2m_0 c^2$ and have created an electron with an energy just over $m_0 c^2$. Where has the energy difference of $m_0 c^2$ gone? We believe that positive energy is also conserved absolutely; it does not disappear into some negative-energy sea.

These problems of interpretation are resolved by proposing that the hole left in the negative-energy sea represents a perceptible, positive-energy particle with an electrical charge opposite to that on the electron. (The absence of a negative-energy particle represents the presence of a positive-energy particle.) This particle is referred to as the *anti-particle* of the electron, is called the *positron*, and is denoted by e^+.

The positron was first discovered in 1931 by

the American physicist Carl Anderson in a cloud chamber photograph of cosmic rays.

Although the arguments given here have concentrated specifically on the electron and the positron, it is important to appreciate that the Dirac equation applies to any relativistic spin-$\frac{1}{2}$ particle, and so too do the ideas of a negative-energy sea and antiparticles. Both the proton p and neutron n can be described by the Dirac equation and seas of negative-energy protons and neutrons may be proposed as coexisting with those of the electrons. The holes in those seas, the antiprotons denoted $\bar{p}$, and antineutrons denoted $\bar{n}$, took somewhat longer to discover than the positron as, in their case, $2m_0c^2$ is large. It requires high-energy accelerators to provide probes that are energetic enough to boost the antiprotons into existence. These were not available until the mid-1950s.

The electron wavefunction which is described in the Dirac equation can now be appreciated in its full four-component form. These components describe, respectively, the spin-up and spin-down states of both the electron and the positron.

The development of the next concept in the microworld is contained in the behaviour of particles and antiparticles. We suggested that an energetic photon can promote a negative-energy electron from the sea, thus leaving a hole. So the photon can create an electron–positron pair from the vacuum. (In fact, this must take place in the presence of another particle to ensure conservation of energy and momentum, see Figure 4.2.) Similarly, an electron and a positron can annihilate each other and give rise to energetic photons. The upshot of this is that particles such as the electron can no

Fig. 4.1. Dirac's energy spectrum of electronic states (*a*) and its interpretation (*b*).

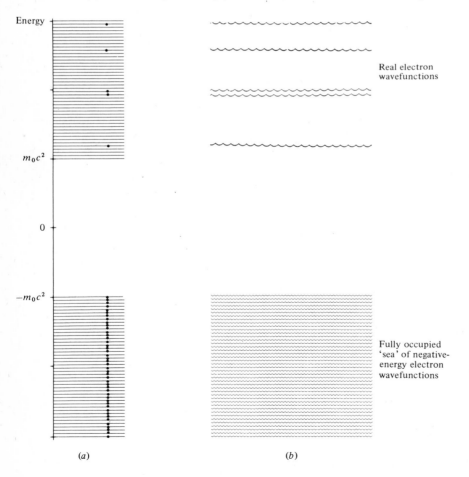

Real electron wavefunctions

Fully occupied 'sea' of negative-energy electron wavefunctions

(*a*)   (*b*)

longer be regarded as immutable, fundamental entities. They can be created and destroyed just like photons, the quanta of the electromagnetic field.

4.4 Quantum field theory

In the most sophisticated form of quantum theory, all entities are described by fields. Just as the photon is most obviously a manifestation of the electromagnetic field, so too is an electron taken to be a manifestation of an electron field and a proton of a proton field. Once we have learned to accept the idea of an electron wavefunction extending throughout space (by virtue of Heisenberg's uncertainty principle for a particle of definite momen-

tum), it is not too great a leap to the idea of an electron field extending throughout space. Any one individual electron wavefunction may be thought of as a particular frequency excitation of the field and may be localised to a greater or lesser extent dependent on its interactions.

The electron field variable is, then, the (Fourier) sum over the individual wavefunctions where coefficients multiplying each of the individual wavefunctions represent the probability of the creation or destruction of a quantum of that particular wavelength (momentum) at any given point. The representation of a field as the summation over its quanta, with coefficients specifying the probabilities of the creation and destruction of those quanta, is referred to as *second quantisation*.

First quantisation is the recognition of the particle nature of a wave or of the wave nature of a particle (the Planck–Einstein and de Broglie

Fig. 4.2. Pair creation by a photon γ in the Dirac picture in (*a*), and in a space-time diagram in (*b*). Energy and momentum conservation require the subsequent involvement of a nearby nucleus.

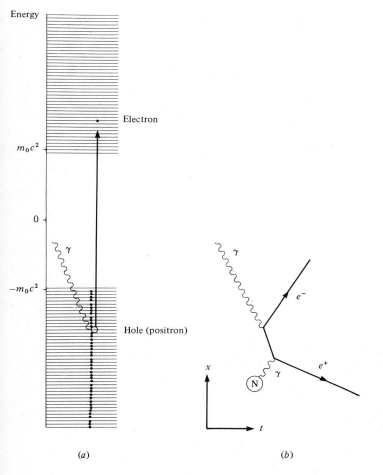

Energy

Electron

$m_0 c^2$

0

γ

$-m_0 c^2$

Hole (positron)

γ

x

γ

(N)

e^-

e^+

t

(*a*)   (*b*)

hypotheses respectively). Second quantisation is the incorporation of the ability to create and destroy the quanta in various reactions.

We are familiar with the electromagnetic and gravitational fields because, their quanta being bosons, there are no restrictions on the number of quanta in any one energy state and so large assemblies of quanta may act together coherently to produce macroscopic effects. Electron and proton fields are not at all evident because, being fermions, the quanta must obey Pauli's exclusion principle and this prevents them from acting together in a macroscopically observable fashion. So although we can have concentrated beams of coherent photons (laser beams), we cannot produce similar beams of electrons. These instead must resemble ordinary incoherent lights (e.g. torchlights) with a wide spread of energies in the beam.

4.5 Interacting fields

Having introduced this new, rather nebulous, concept of a field representation of matter, we must now set about using it. Our ultimate objective must be to predict the values of physical quantities which can be measured in the laboratory such as particle reaction cross-sections, particle lifetimes, energy levels in bound systems, etc. We hope to achieve this by using the idea of quantum fields to tell us the probabilities of the creation and destruction of their quanta in various reactions, and to provide us with descriptions of the behaviour of the quanta between creation and destruction (the wavefunctions). This will then allow us to calculate the probabilities associated with physical processes.

To formulate this, we must be guided by some principle from which all mechanics can be derived. In fact, the same principle, Hamilton's variational principle, can be used to derive Newtonian mechanics, quantum mechanics and quantum field theory. The *Lagrangian L* for any system is the difference between its kinetic energy (*KE*) and its potential energy (*PE*).

$$L = KE - PE.$$

For a classical particle, say a cricket ball, moving through the gravitational field of the earth, the potential energy is due to its height x above the earth ($PE = mgx$), and its kinetic energy is due to its velocity ($KE = \frac{1}{2}mv^2$).

Hamilton's principle states that the evolution of any system is such as to minimise L. So the path of the cricket ball is dictated by those values of position x and velocity v for which L remains at its minimum value. Put another way around, by minimising L with respect to x and v we can obtain the equations of motion of the cricket ball.

$$\delta L(x,v) = 0 \text{ gives } F = ma.$$

This entirely general principle can also be used in quantum mechanics. (In the quantum version, however, as we are dealing with wavefunctions (or, more properly, fields) which extend throughout space, we do not deal with the total Lagrangian L directly, but with the Lagrangian density $\mathscr{L}$. The total Lagrangian can then be found by integrating the Lagrangian density over all space. Although in future discussions we shall be talking about the properties of the Lagrangian, the comments will properly apply to the Lagrangian density, a fact which we will acknowledge by using the symbol $\mathscr{L}$.)

It is possible to write down the expression for the Lagrangian density of a free electron as a function of the free electron wavefunction. By minimising $\mathscr{L}$ with respect to the wavefunction and its time and space derivatives, it is possible to derive Dirac's equation of motion of the electron, denoted $D\psi_e$.

$$\delta \mathscr{L}(\psi_e) = 0 \text{ gives } D\psi_e.$$

In both classical and quantum mechanics it was, of course, the equations of motion which were discovered first.

We regress to the Lagrangian and work forward from there for the sake of generality. But in the case of elementary particles *in interaction* we do not know in general the equations of motion and, where we do, we cannot solve them. We cannot therefore proceed to calculate the quantities of physical interest resulting from the motions of particles.

4.6 Perturbation theory

To describe elementary particle reactions in which quanta can be created and destroyed, it is necessary to propose an expression for the Lagrangian of the interacting quantum fields. Let us concentrate on interacting electron and photon fields only. The Lagrangian will contain parts which represent free electrons $\mathcal{L}_0(\psi_e)$ and free photons $\mathcal{L}_0(A)$, where A denotes a four-vector representing the electromagnetic field. It will also contain parts which represent the interactions between electrons and photons, $\mathcal{L}_{INT}(\psi_e, A)$, whose form will be dictated by general principles. These will include, for instance, Lorentz invariance and various conservation laws which the interactions are observed to respect (such as the conservation of electrical charge). In Chapter 11 we will see how these principles can be expressed in terms of the symmetry of the Lagrangian under various groups of transformations.

The total Lagrangian is then the sum of all these parts,

$$\mathcal{L} = \mathcal{L}_0(\psi_e) + \mathcal{L}_0(A) + \mathcal{L}_{INT}(\psi_e, A).$$

This is the top-level specification of the fields being described and the way in which they interact. We can proceed to predict the values of physical quantities by carrying out a modern version of Hamilton's variational principle. Instead of giving the equations of motion for free electrons and photons, the equations are now modified by the existence of the interactions. The variational principle now describes the propagation of the fields in terms of the probabilities of the creation and the destruction of the quanta of the fields and by the wavefunctions of the quanta during their existence (referred to as propagators in this context).

In the late 1940s the American physicist Richard Feynman derived a set of rules which specifies the propagation of the fields as the sum of a set of increasingly complicated sub-processes involving the quanta of the interacting fields. Each sub-process in the sum can be represented in a space–time diagram referred to as a Feynman diagram. The rules associate with each diagram a mathematical expression describing the wavefunctions of the particles involved. To calculate the probability of occurrence P of any physical event involving the quanta of the fields, it is first necessary to specify the initial and final states being observed, denoted $|i\rangle$ and $\langle f|$ respectively, and then to select all the Feynman diagrams which can connect the two. The mathematical expression for each diagram is then worked out: essentially, the wavefunctions of the quanta are multiplied together to give the quantum-mechanical amplitude m for the sub-pro-

cess. The amplitude for a number of the individual sub-processes may then be added to give the total amplitude M which is then squared to give the required probability of occurrence.

$$P = |\langle f | M | i \rangle|^2,$$
$$M = m_1^{(1)} + m_2^{(1)}$$
$$m_1^{(2)} + m_2^{(2)} + m_3^{(2)} + \ldots.$$

In this notation $m_i^{(1)}$ denotes the 'first-order' diagrams with just two photon–electron vertices involved, $m_i^{(2)}$ denotes 'second-order' diagrams with four photon–electron vertices, $m_i^{(3)}$ denotes 'third-order' diagrams and so on.

For example, in the case of electron–positron elastic scattering, the initial and final states are $|e^+e^-\rangle$ and $\langle e^+e^-|$ respectively. A few of the simplest Feynman diagrams connecting the two are shown in Figure 4.3. The first sub-process, amplitude $m_1^{(1)}$, is the exchange of a photon between the electron and the positron; the second, $m_2^{(1)}$, is the annihilation of the electron and the positron into a photon and its subsequent reconversion; the third $m_1^{(2)}$, is the exchange of two photons, and so on.

The probability of occurrence (i.e. of the transformation between initial and final states) may

Fig. 4.3. The perturbation series containing the various sub-processes possible in electron–positron scattering.

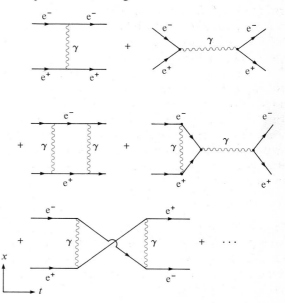

then be restated as the cross-sectional area of two colliding particles, as the mean lifetime for a particle to decay, or as some other appropriate measurable parameter. This is achieved by adopting the kinematical prescriptions which take into account factors like the initial flux of colliding particles, the density of targets available in a stationary target and so on.

The reason why this approach can be adopted is that only the first few of the simplest Feynman diagrams from the infinite series need be considered. This is because the energy of the interaction between electrons and photons (the strength of the electromagnetic force), is small compared with that of the free particles. It can be regarded as a perturbation to free particle-type behaviour. Another way of stating this is that the probability of the electron or positron interacting with a photon is small. In fact, each photon–electron vertex multiplies the probability of occurrence of the diagram by $e/\sqrt{(\hbar c)}$. As each new order of diagram contains a new photon line with two vertices, the relative magnitude of successive orders is reduced by $e^2/(\hbar c) = \frac{1}{137}$. So only the first few sub-processes need be calculated to achieve an acceptable approximation to the exact answer.

Summary	
The Lagrangian	specifies the form of the interaction between the fields.
The Variational Principle	provides the equations of motion from L.
The Perturbation Principle	approximates the equations of motion by a series of …
Feynman Diagrams	which show sub-processes between initial and final states involving quanta which may be calculated to give …
Probabilities of Physical Events	which may be stated as cross-sections, lifetimes, etc.

4.7 Virtual processes

It is important to understand that the dynamics of the individual field quanta within any sub-process of the perturbation expansion are *not* constrained by energy or momentum conservation, provided that the sub-process as a whole does conserve both. This microscopic anarchy is permitted by Heisenberg's uncertainty principle which states that energy can be uncertain to within ΔE for a time Δt, such that

$$\Delta E\, \Delta t \geqslant \hbar.$$

So an electron may emit an energetic photon, or a photon may convert into an electron–positron pair over microscopic timescales, provided that energy conservation is preserved in the long run.

These illicit processes are known as 'virtual processes'. They form the intermediate states of elementary particle reactions. So although we do not see them, we must calculate the probabilities of their occurrence and add them all up to find the number of different ways for a particle reaction to get from its initial to its final state. A good example of a virtual process is the annihilation of an e^+e^- pair into a photon. The energy of the e^+e^- pair must be

$$E_{e^+e^-} = (m_e^2 c^4 + p_e^2 c^2)^{\frac{1}{2}} + (m_e^2 c^4 + p_{e^-} c^2)^{\frac{1}{2}},$$

whereas the energy momentum relation of the photon is

$$E_\gamma = p_\gamma c.$$

So it is not possible to have both

$$E_{e^+e^-} = E_\gamma \text{ and } p_\gamma = p_{e^+} + p_{e^-}$$

because of the rest mass of the e^+e^- pair. This means that the virtual photon can exist only as an unobservable intermediate state before dissolving into a collection of material particles which do conserve energy and momentum.

4.8 Renormalisation

In writing down all the Feynman diagrams of the sub-processes we find some whose amplitude (product of wavefunctions) appears to be infinite. These diagrams are generally those with bubbles on either electron or photon wavefunctions or surrounding electron–photon vertices, see Figure 4.4.

Fig. 4.4. (*a*) (*b*) and (*c*) Diagrams with 'bubbles' which give infinite contributions to the perturbation series.

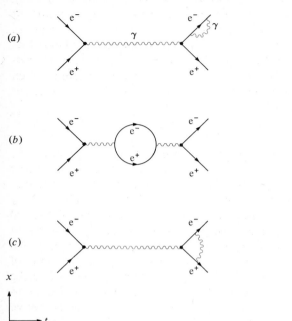

(*a*)

(*b*)

(*c*)

Fig. 4.5. (*a*) The completed ('dressed') electron wavefunction already contains its quantum corrections (interactions with virtual photons). (*b*) The photon propagation likewise.

These diagrams give infinite contributions due to ambiguities in defining the electron and the photon.

An ordinary electron propagating through space is constantly emitting and absorbing virtual photons. It is enjoying self-interaction with its own electromagnetic field (of which its own charge is the source). So the wavefunction of the electron is already dressed up with these virtual photons, see Figure 4.5(*a*). Similarly, a photon propagating through space is free to exist as a virtual e^+e^- pair, and the full photon wavefunction already contains the probabilities of this occurring, Figure 4.5(*b*). Also, the electric charge, which we denote e, already contains the quantum corrections implied by the diagram of Figure 4.4(*c*).

In 1949, Feynman, Dyson & Tomonaga showed how the infinite contributions to the perturbation series can be removed by redefining the electron, photon and electric charge to include the quantum corrections. When the real electrons, photons and charges appear, the infinite diagrams are included implicitly and should not be recounted. The mathematical proof of this demonstration is known as 'renormalisation'.

Renormalisation is a necessary formal process which shows that the particles in the theory and their interactions are consistent with the principles of quantum theory. These may seem like hollow words for the familiar interactions of electrons with photons. But in the more esoteric quantum field theories we are going to encounter, both particle

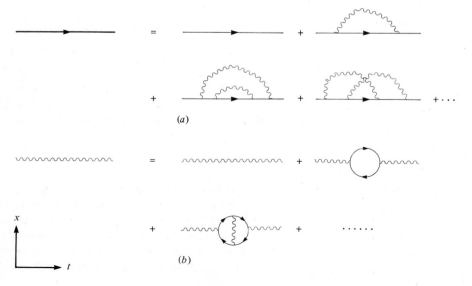

content of the theories and the form of their interactions are largely unknown. In these cases, the ability to renormalise the perturbation expansion of the Lagrangian is a good guide to the acceptability of the theory.

4.9　Quantum electrodynamics

This is the name (often abbreviated to QED) given to the relativistic quantum field theory describing the interactions of electrically charged particles via photons. The discovery of the perturbation expansion reveals the existence of an infinite number of ever-decreasing quantum corrections to any electromagnetic process. The renormalisability of QED means that we can avoid apparently infinite contributions to the perturbation expansion by careful definition of the electron and photon. Therefore we can calculate the value of observable parameters of electromagnetic processes to any desired degree of accuracy, being limited only by the computational effort required to evaluate the many hundreds of Feynman diagrams which are generated within the first few orders (first few powers of $e^2/(\hbar c)$) of the perturbation expansion. This has lead to some spectacular agreements between theoretical calculations and very accurate experimental measurements.

The g factor of the electron is not, in fact, exactly equal to 2 (as predicted by the Dirac equation). Its value is affected by the quantum corrections to electron propagator illustrated in Figure 4.5(*a*). Essentially, the virtual photons of the quantum corrections carry off some of the mass of the electron whilst leaving its charge unaltered. This can then affect the magnetic moment generated by the electron during interactions. The measure of agreement between QED and experimental measurement is given by the figure for the modified g factor.

$$\frac{g}{2} = \pm \begin{array}{l} 1.\,001\,159\,652\,41 \\ 0.\,000\,000\,000\,20 \end{array} \quad \begin{array}{l}\text{experimental}\\\text{measurement,}\end{array}$$

$$\frac{g}{2} = \pm \begin{array}{l} 1.\,001\,159\,652\,38 \\ 0.\,000\,000\,000\,26 \end{array} \quad \begin{array}{l}\text{theoretical}\\\text{prediction.}\end{array}$$

There are several other such amazing testaments to the success of QED, including numbers similar to the above for the g factor of the muon (a heavy brother of the electron which we will meet soon), and yet more subtle shifting of the exact values of the energy levels within the hydrogen atom, the so-called Lamb shift.

This success makes QED the most precise picture we have of the physical world (or at least the electromagnetic phenomena in it). For this reason we shall look at QED again in Part 6 in an attempt to discover the fundamental principles behind it (i.e. behind the form of the interaction between the fields). This is so that we can attempt to repeat the theory's success for the other forces in nature.

4.10　Postscript

We have now looked at the frontiers of physics as they appeared at the turn of the century and have seen that relativity and quantum mechanics emerged in turn from the vacuum of knowledge beyond those frontiers. The realisation that relativity and quantum mechanics must be made mutually consistent has led to the discovery of antiparticles, which leads in turn to the concept of quantum fields. The theory of interacting quantum fields is the most satisfactory description of elementary particle behaviour. All calculations in quantum field theory follow from the specification of the correct interaction Lagrangian, which is determined by the conservation laws obeyed by the force under study.

We have developed this picture of the world almost exclusively in terms of the particles interacting by the electromagnetic force. It is now time to turn our attention to the other particles and forces in nature to see if they are amenable to a similar treatment.

In what follows, we will not introduce the language of quantum fields again. We will deal only in terms of particle wavefunctions where necessary. This is perfectly acceptable as far as we are concerned, indeed as far as most high-energy physicists are concerned, because once the Feynman rules for any theory have been derived, the origin of the wavefunction in the fields is of largely historical interest. But we will be examining the symmetry of Lagrangians in new gauge theories. Although we shall speak roughly of them containing only wavefunctions, we should appreciate that, properly, the arguments apply to the underlying fields.

Part 1
Basic particle physics

5

The fundamental forces

5.1   Introduction

It is an impressive demonstration of the unifying power of physics to realise that all the phenomena observed in the natural world can be attributed to the effects of just four fundamental forces. These are the familiar forces of gravity and electromagnetism, and the not-so-familiar weak and strong nuclear forces (generally referred to as the 'weak and strong forces'). Still more impressive is the fact that the phenomena occurring in the everyday world can be attributed to just two: gravity and electromagnetism. This is because only these forces have significant effects at observable ranges. The effects of the weak and strong nuclear forces are confined to within, at most, 10^{-15} m of their sources.

With this in mind, it is worthwhile summarising a few key facts about each of the four forces before going on to look at the variety of phenomena which they display in our laboratories. In each case we are interested in the sources of the force and the intrinsic strength of the interactions to which they give rise. We are interested also in the space–time properties of the force: how it propagates through space and how it affects the motions of particles under its influence. Finally, we must consider both the macroscopic (or classical) description of the forces (where appropriate) and the microscopic (or quantum-mechanical) picture (where possible).

5.2　Gravity

Gravity is by far the most familiar of the forces in human experience, governing phenomena as diverse as falling apples and collapsing galaxies. The source of the force is mass and, because there is no such thing as negative mass, the force is always attractive between two masses and is always directed along a line joining the two. The force is independent of all other attributes of the bodies acted upon, such as their electric charges, spins, directions of motions, etc.

The gravitational force is described classically by Newton's famous inverse square law which states that the magnitude of the force between two particles is proportional to the product of their masses and inversely proportional to the square of the distance between them.

$$F = G \frac{m_1 m_2}{r^2}.$$

The strength of the force is governed by Newton's constant G and is extremely feeble compared to the other forces, see Table 5.1. We notice the effects of gravity only because it is the *only* long-range force acting between electrically neutral matter. In the microworld the effects of gravity are mainly negligible and it is only our desire for completeness which leads us to include it in the context of elementary particle physics. Only in esoteric regions, such as on the boundary of a black hole, do the effects of gravity on the elementary particles become important.

The mechanism which gives rise to this force in the classical picture is that of the gravitational field which spreads out from each mass-source to infinity. A test mass will interact not with the mass-source directly (which would require instantaneous action at a distance, a philosophically unattractive concept), but with the gravitational field. At each point in space, this carries the knowledge of its parent's mass and of what potential to offer the test mass (i.e. what force to exert on it).

Another feature of Newton's formula is that the property it uses to characterise the sources (their mass) is identical to the property which characterises their response to accelerations, as given by another famous formula of Newton's:

$$F = ma.$$

This equivalence between gravitational and inertial mass led Einstein to speculate on the identity of the effects of gravity with those of acceleration. This is the principle of equivalence which forms the starting point of general relativity.

5.2.1　*General relativity*

We saw how in special relativity, observers' perceptions of time and space were dilated by factors depending on their relative velocities. From this it follows that during acceleration (changing velocity) an observer's scales of time and space must distort to match up with the formula of special relativity. By the principle of equivalence, an acceleration is identical to the effects of a gravitational field, so this too must give rise to a distortion of space–time. Einstein's general relativity goes on to explain the somewhat tenuous reality of the gravita-

Table 5.1. *Relative strengths of forces as expressed in natural units*

Force	Range	Strength	Acts on
Gravity	∞	$G_{\text{Newton}} \approx 6 \times 10^{-39}$	All particles
Weak nuclear force	$< 10^{-18}$ m	$G_{\text{Fermi}} \approx 1 \times 10^{-5}$	Leptons, hadrons
Electromagnetism	∞	$\alpha = \frac{1}{137}$	All charged particles
Strong nuclear force	$\approx 10^{-15}$ m	$g^2 \approx 1$	Hadrons

The dimensionality of the forces is removed by dividing out the appropriate powers of $\hbar$ and c, to leave a dimensionless measure of the forces' intrinsic strengths

tional field as the warping of space–time around a mass-source. Thus a mass will distort space–time rather like a bowling ball laid on a rubber sheet. And the distortion due to gravity of the trajectory of a test mass passing the mass will be analogous to rolling a marble around the curve of the rubber sheet, see Figure 5.1.

The theory goes on to predict the existence of *gravitational waves* propagating through space as the result of some changes in a mass-source (such as the collapse of a conventional star into a neutron star or a black hole). In this event, distortions in space–time will spread out spherically in space at the speed of light, rather like ripples spread out circularly across the surface of a still pond into which a stone is dropped. There has been considerable experimental effort devoted to attempts to detect such waves resulting from cosmic events out in space, but none have so far succeeded, most probably because such disturbances are too small to affect noticeably our most sensitive detectors.

5.2.2 *Quantum gravity*

It is important to remember that Einstein's general relativity is still a classical theory, it does not account for gravity in the quantum-mechanical regime. A successful quantum theory of gravity has not yet been formulated, and the reconciliation of general relativity with quantum mechanics is one of the major outstanding problems in theoretical physics. It is straightforward enough to take the first few steps towards such a theory, following an analogy with quantum electrodynamics. We can propose that the gravitational field consists of microscopic quanta called *gravitons* which must be massless (to accommodate the infinite range of gravity) and of spin 2 (necessary to produce only an attractive force). The gravitational force between any two masses can then be described as an exchange of gravitons between the two. Problems arise, however, because, unlike quantum electrodynamics, certain graviton sub-processes always seem to occur with an infinite probability (quantum gravity is not renormalisable). The problem is compounded by the high spin of the graviton, which makes calculations difficult, and by the distortions of space–time induced by their propagation. These

Fig. 5.1. According to general relativity, mass bends space–time, which gives rise to the trajectories associated with gravity.

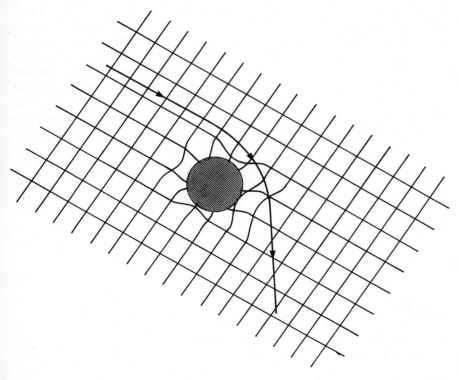

problems remain unsolved, but some of the most recent 'super-symmetric' theories at least address them in an encouraging fashion, see Part 9.

Another hinderance preventing progress in quantum gravity is the total absence of observable microscopic gravitational phenomena for experiments. Not only are they masked by the other three forces, but they are estimated to be many orders of magnitude below any thresholds of detectability even if they were not. It is as if we were being asked to derive the theory of QED from the behaviour of pith-balls and magnets!

5.3   Electromagnetism

This is the force of which we have the fullest understanding despite its rather complicated nature. This is possibly a reflection of its physical characteristics: it is of infinite range, allowing macroscopic phenomena to guide our understanding of classical electromagnetism, and it is a reasonably strong force, allowing its microscopic phenomena to be observed and to guide our formulation of QED.

The source of this force is, of course, electric charge which can be either positive or negative, leading to an attractive force between unlike charges and a repulsive force between like charges. When two charges are at rest, the electrostatic force between them is given by Coulomb's law, which is very similar to Newton's law of gravity, namely that the magnitude of the force is proportional to the product of the magnitudes of the charges involved (empirically observed to exist only as multiples of the charge on an electron) and inversely proportional to the square of the separation between them.

$$F = K \frac{N_1 e \cdot N_2 e}{r^2},$$

where N_1 and N_2 are integral multiples of the charge on the electron e and K is a constant depending on the electrical permittivity of free space. New mysteries are introduced by the concept of electric charge. What is it, other than a label for the source of a force we observe to act? Why does it exist only in quanta? Why is the charge quantum on the electron exactly opposite to that on the proton? These are largely taken for granted in classical electromagnetism and are only now being addressed in the modern theories described in Part 9.

Unlike gravity, when electric charges start to move, qualitatively new phenomena are introduced. A moving charge has associated with it not only an electric field, but also a magnetic field. A test charge will always be attracted (or repelled) along the direction of the electrical field, i.e. along a line joining the centres of the two charges. But the effect of the magnetic field is that a test charge will be subjected to an additional force along a direction which is mutually perpendicular to the relative motion of the source charge and to the direction of the magnetic field, see Figure 5.2. These properties imply that the combined electromagnetic force on a particle cannot be described simply by a number representing the magnitude of the force but, instead, by a vector quantity describing the magnitude of the forces acting in each of the three directions.

When a charge is subject to an acceleration, then a variation in electric and magnetic fields is propagated out through space to signal the event. If it is subject to regular accelerations, as may occur when an alternating voltage is applied to a radio aerial, then the charge emits an electromagnetic wave which consists of variations in the electric and magnetic fields perpendicular to the direction of propagation of the wave, see Figure 5.3. Such an electromagnetic wave is part of the electromagnetic spectrum which contains, according to the frequency of oscillation of the fields, radio waves, infra-red waves, visible light, ultra-violet light, X-rays and gamma rays. See Figure 5.4.

Electromagnetic phenomena are all described

Fig. 5.2. The motion of a charged particle in a magnetic field directed out of the plane of the paper.

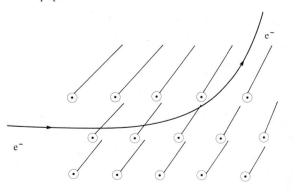

in the classical regime by Maxwell's equations which allow us to calculate, say, the electric field resulting from a particular configuration of charges or the wave equation describing the propagation of electric and magnetic fields through space.

One interesting feature of these equations is that they are asymmetric owing to the absence of a fundamental quantum of magnetic charge. It is possible to conceive of a source of a magnetic field which would give rise to an elementary magnetostatic force. Such a magnetic charge would appear as a single magnetic pole, in contrast to all examples of terrestial magnets which consist invariably of north-pole–south-pole pairs. These conventional magnets are not fundamental, but are the result of the motions of the atomic electrons. The possibility of the existence of truly fundamental magnetic monopoles has been revived recently following their emergence from the most modern theories and reported sightings, see Part 9.

We have already seen, in Part 0, how we can

Fig. 5.3. The propagation of an electromagnetic wave resulting from the regular accelerations of a charge.

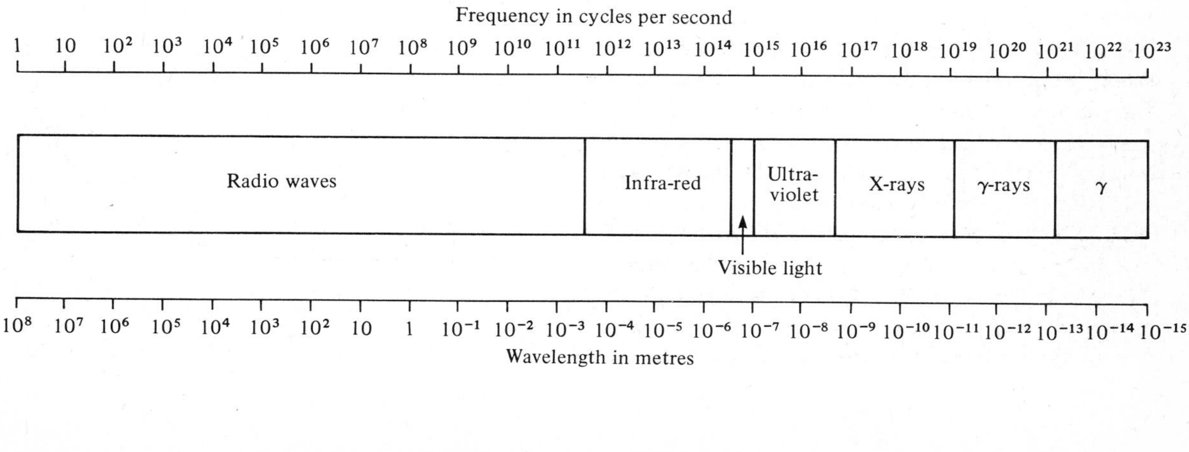

formulate the quantum theory of electrodynamics by describing the interactions of charged particles via the electromagnetic field as the exchange of the quanta of the field, the photons, between the particles involved. QED is the paradigm quantum theory towards which our descriptions of the other forces all aspire.

5.4 The strong nuclear force

When the neutron was discovered by James Chadwick in 1932, it became obvious that a new force of nature must exist to bind together the neutrons and protons (referred to generically as *nucleons*) within the nucleus. (Prior to this discovery physicists seriously entertained the idea that the nucleus might have consisted of protons and electrons bound together by the electromagnetic force.) Several features of the new force are readily apparent.

Firstly, as the nucleus was realised to consist only of positively charged protons and neutral neutrons confined within a very small volume (typically of diameter 10^{-15} m), the strong force must be very strongly attractive to overcome the intense mutual repulsions felt by the protons. The

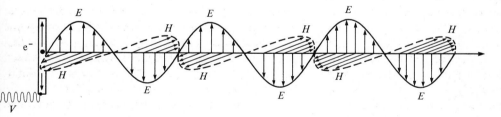

Fig. 5.4. The electromagnetic spectrum.

binding energy of the strong force between two protons is measured in millions of electron volts, MeV, as opposed to typical atomic binding energies which are measured in electron volts (see Appendix 1 for definitions of these units of energy).

The second fact concerning the strong nuclear force is that it is of extremely short range. We know that the electromagnetic force alone can explain the arrangement of electronic orbits within atoms (corresponding typically to radii of 10^{-10} m), and that Rutherford's early scattering experiments of α particles through atoms could be described by the electromagnetic force alone. Only at higher energies, when the α particles are able to approach the nuclei more closely, are any effects of the strong force found. In fact, the force may be thought of as acting between two protons only when they are actually touching, implying a range of the strong force similar to that of a nuclear diameter of about 10^{-15} m.

Finally, the last fact we shall mention is that the strong nuclear force is independent of electric charge in that it binds both protons and neutrons in a similar fashion within the nucleus.

One consequence of the solely microscopic nature of the strong force is that we should expect it to be a uniquely quantum phenomenon. We can expect no accurate interpretation in terms of classical physics but only in the probabilistic laws of quantum theory.

One of the prime sources of early information on the strong force was the phenomenon of radioactivity and the question of nuclear stability. This involves the explanation of the neutron/proton ratios of the stable or nearly stable nuclei. These can be displayed as a band of stability on the plane defined by the neutron number, N, and the proton number, Z, of the nucleus, see Figure 5.5

The fact that it is predominantly the heavier nuclei which decay confirms our picture of the short-range picture of the strong force. If we naively think of the nucleus as a bag full of touching spheres, then if the force due to any one nucleon source were able to act on all other nucleons present, we would expect nuclei with more nucleons to enjoy proportionately stronger binding and thus greater stability. (Adding the nth nucleon to a nucleus would give rise to an extra $(n-1)$ nuclear bonds and so a binding energy which increases with

n.) This is observed not to be the case. It is the heavier nuclei which suffer radioactive decay, indicating an insufficient binding together of the nucleons. This is because the nuclear force acts only between touching, or 'nearest-neighbour' nucleons. The addition of any extra nucleon will then give rise only to a constant extra binding energy whereas the electric repulsion of the protons is a long-range force and does grow with the number of protons present.

Thus the question of nuclear stability may be described in part by the balance of the repulsive electrical forces and the attractive strong forces affecting the nucleons. It is possible to calculate the sum of these two forces for each nucleus and so to calculate the average binding energy per nucleon in each case, the more negative the binding energy indicating that the more strongly bound are the nucleons within the nuclei. This can be shown graphically as in Figure 5.6. The relatively small negative binding energy of the light atoms results from them not having enough nucleons to saturate fully the nearest-neighbour strong interactions available. The most strongly bound nuclei are those

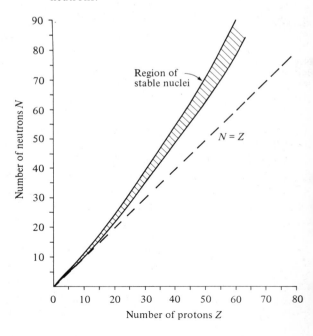

Fig. 5.5. Nuclear stability against radioactive decay is governed by the ratio of protons to neutrons.

in the mid-mass range, like iron, which more efficiently use the strong force without incurring undue electric repulsion. The heavier nuclei suffer because the electric repulsion grows by an amount proportional to the number of protons present.

As nature attempts to accommodate heavier and heavier nuclei, a point is reached where it becomes energetically more favourable for the large nuclei to split into two more-tightly bound, mid-mass-range nuclei. This gives rise to an upper limit on the weights of atoms found in nature, occupied by uranium[238] with 92 protons and 146 neutrons. By bombarding uranium with neutrons, it is possible to exceed nature's stability limit causing the uranium + neutron nucleus to split into two. This is nuclear fission.

Radioactive α decay occurs when an element is not big enough to split into two, but would still like to shed some weight to move down the binding energy curve to a region of greater stability. The α particle (which is a helium nucleus consisting of two protons and two neutrons) will have existed as a 'nucleus within a nucleus' prior to the decay. By borrowing energy for a short time according to Heisenberg's uncertainty principle, it will be able to travel beyond the range of the strong attractive forces of the remaining nucleons to a region where it

is subject only to the electrical repulsion due to the protons. Thus the nucleus is seen to expel an α particle, see Figure 5.7. Because the energy is borrowed according to the probabilistic laws of quantum theory, it is not possible to specify a particular time for α decay, but only to specify the time by which there will be a, say, 50% probability of a given nucleus having undergone the decay (corresponding to the average time needed for 50% of a sample to decay). This is called the half-life of the element, denoted by τ.

Another feature of nuclear stability can be explained by the action of another quantum principle. Although we have explained why too many protons in a massive nucleus may cause it to break up, we have not explained why this cannot be countered by simply adding an arbitrary number of neutrons to gain extra attractive strong forces. The reason is due to Pauli's exclusion principle. Because both protons and neutrons are fermions, no two protons and no two neutrons can occupy the same quantum state. We cannot simply add an arbitary number of neutrons to dilute the repulsive effects of the electric charges on the protons, as the exclusion principle forces the neutrons to stack up in increasingly energetic configurations leading to a reduction in the negative binding energy per nucleon and so to decreased stability.

Fig. 5.6. Nuclear stability can be expressed in terms of the binding energy available per nucleon in the nucleus.

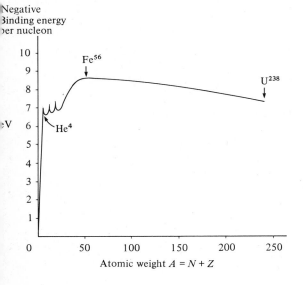

Fig. 5.7. Radioactive α decay. An α particle within the nucleus borrows enough energy via Heisenberg's uncertainty principle to overcome the potential binding energy of the strong force.

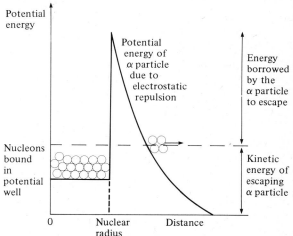

Although we have now reviewed some facts about the strong force (its short range, charge independence and spin dependence via the exclusion principle, etc.), we have done nothing to explain the mechanism of its action apart from noting that only a quantum picture will be suitable for such a microscopic phenomenon. Yukawa's formulation of his meson theory of the strong force is the point of departure into particle physics proper from the inferences of nuclear physics just discussed, and is described in Chapter 12.

5.5   The weak nuclear force

One of the most obvious features of the neutron is that it decays spontaneously into a proton and an electron with a half-life of about 18 minutes. This period is much longer than any of the phenomena associated with the strong force, and there is no reason to expect the influence of the electromagnetic force to affect the uncharged neutron, so it is clear that neutron decay is due to some qualitatively new force of nature.

It is this weak force causing neutron decay which lies behind the phenomenon of the radioactive β decay of nuclei described in Chapter 1. The decay of the neutron into a proton allows a nucleus to relieve a crucial neutron surplus which, because of the action of the Pauli exclusion principle, may be incurring a substantial energy penalty and eroding the binding energy of the nucleus.

The same interaction may also allow the reverse reaction to occur in which a nuclear proton transfers into a neutron by absorbing an electron. (This may occur because of the very small but finite chance that the electron may find itself actually inside the nucleus, according to the positional uncertainty represented by the electron wavefunction.) This reaction will allow a proton-rich nucleus suffering from undue electric repulsion to dilute its proton content slightly, thereby strengthening its binding.

One problem soon encountered in attempts to explain radioactive β decay is that the electrons which are emitted from decaying nuclei are seen to emerge with a range of energies up to some maximum which is equal to the difference in the masses of the initial and final nuclei involved. When the electrons emerge with less than this maximum figure we seem to have lost some energy. To avoid

this apparent violation of energy conservation (and also an accompanying apparent violation of angular momentum conservation), Pauli postulated in 1930 that another, invisible, particle was also emitted during the decay, which carries off the missing energy and angular momentum. As the original reactions do conserve electric charge, then this new particle must be neutral. On the strength of this, Fermi called it the *neutrino*.

Several properties of the neutrino are apparent from the facts of β decay. To restore conservation of energy, it is necessary that the neutrino be massless (because some electrons do emerge with the maximum energy allowed by the mass difference). Similarly, to restore angular momentum conservation the neutrino must be spin $\frac{1}{2}$. From these two facts, relativity tells us that the masslessness of the neutrino allows it to exist only when travelling at the speed of light and this in turn allows it to exist with a definite spin only. A neutrino cannot flip its spin orientation as can, say, the massive spin-$\frac{1}{2}$ electron. Another interesting feature is that the neutrino interacts with other particles only by the weak force and gravity (because the strong interaction is obviously not present in neutron decay, and because the uncharged neutrino experiences no electromagnetic effects).

The apparent invisibility of the neutrino is due to the very feebleness of the weak force, as indicated in Table 5.1. This reluctance to interact allows it to pass through the entire mass of the earth with only a minimal chance of interaction en route. Because of this, the neutrino was not observed (i.e. collisions attributable to its path were not identified) until large neutrino fluxes emerging from nuclear reactors became available. This was achieved in 1956 by Reines, some 26 years after Pauli's proposal.

The weak force, like its strong counterpart, acts over microscopic distances only. In fact, to all intents and purposes it makes itself felt only when particles come together at a point (i.e. below any resolving power available to physics, say less than 10^{-18} m). This allows its description in terms of quantum physics only, which we shall discuss further in Parts 3, 4 and 5.

6

Symmetry in the microworld

6.1 Introduction

In the everyday world, symmetries in both space and time have a universal fascination for the human observer. In nature, the symmetry exhibited in a snowflake crystal or on a butterfly's wings might be taken to indicate some divine guiding hand, whilst in art the pleasures of design or of a fugue may be seen as its imitation. Pleasing as symmetry may be, however, its significance generally remains unappreciated.

In the world of physics, and especially in the microworld, symmetries are linked closely to the actual dynamics of the systems under study. They are not just interesting patterns or an artistic disguise for science's passion for classification. Indeed, it is no exaggeration to say that symmetries are the most fundamental explanation for the way things behave (the laws of physics).

Historically this has not always been appreciated. It is, of course, the case that physicists notice natural phenomena and write down equations of motion to describe them (notably Newton, Einstein and Dirac, to name but an illustrious few). But in describing the microworld it is generally far too difficult to write down the equations of motion straightaway, the forces are unfamiliar and our experiments provide only groundfloor windows into the skyscraper of the high-energy domain. So we are forced to consider first the symmetries governing the phenomena under study, generally indicated by the action of conservation rules of one

sort or another (e.g. energy, momentum, electric charge). The symmetries may then guide our investigation of the nature of the forces to which they give rise.

Symmetry is described by a branch of mathematics called 'group theory'. A group is simply a collection of elements with specific interrelationships defined by group transformations. The notion is made non-trivial by the demand that repeated transformations between elements of the group are equivalent to another group transformation from the initial to the final elements. When a symmetry group governs a particular physical phenomenon (if the Lagrangian governing the phenomenon does not change under the group transformations), then this implies the existence of a conserved quantity. This connection is due to a mathematical proof called Noether's theorem which states that, for every continuous symmetry of a Lagrangian, there is a quantity which is conserved by its dynamics. This quantity is given by the generator of the group.

We can proceed to put some flesh on this theoretical skeleton by examples of four kinds of symmetry used extensively in particle physics: continuous space–time symmetries, discrete symmetries, dynamical symmetries, and internal symmetries. After a look at each of these, we will mention in closing how even broken symmetries can provide a useful guide to the formulation of physical laws.

6.2 Space–time symmetries

Foremost amongst these are the operations of translation through space, translation through time and rotation about an axis. The physical laws governing any process are formulated with a particular origin and a particular coordinate system in mind; for instance, laws of terrestial gravity might use the centre of the earth as their origin, whilst the laws of planetary motion might use the centre of the sun.

However, the physical laws should remain the same whatever the choice and so any mathematical expression of the laws should remain the same under any of these transformations. Application of Noether's theorem, then, reveals the conserved quantity corresponding to each particular invariance. Invariance under a translation in time (e.g. the laws of physics predict the same evolution of identical processes regardless of when they occur) implies that conservation of energy is built into the laws describing the process. Invariance under a translation in space (e.g. physics is the same in London as in New York) implies conservation of momentum. And invariance under spatial rotations implies conservation of angular momentum, see Figure 6.1.

In practice, all these different invariances are ensured by requiring that the equations describing any system are covariant under the group of Lorentz transformations of special relativity. This can be displayed automatically by writing the equations in a purely four-vector form.

6.3   Discrete symmetries

The continuous space–time symmetries are called *proper* Lorentz transformations because they can be built up from a succession of infinitesimally small ones. However, there are also *improper* symmetries which cannot be so built up. These improper, or discrete, symmetries do not have corresponding conservation laws as important as those of the continuous symmetries. However, they have proved very useful in telling us which particle reactions are possible with a given force and which are not. We shall deal with the three most important improper symmetries in turn.

Fig. 6.1. Rotations redefine a coordinate system. Invariance of the laws of physics with respect to such a rotation implies the conservation of angular momentum.

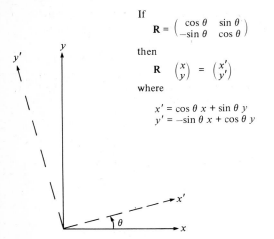

If

$$\mathbf{R} = \begin{pmatrix} \cos\theta & \sin\theta \\ -\sin\theta & \cos\theta \end{pmatrix}$$

then

$$\mathbf{R} \begin{pmatrix} x \\ y \end{pmatrix} = \begin{pmatrix} x' \\ y' \end{pmatrix}$$

where

$$x' = \cos\theta\, x + \sin\theta\, y$$
$$y' = -\sin\theta\, x + \cos\theta\, y$$

6.3.1   *Parity or space inversion*

In this operation, denoted **P**, the system under consideration (e.g. a particle wavefunction) is reflected through the origin of the coordinate system. An alternative way of thinking of this is as the reversal of a right-handed coordinate system into a left-handed coordinate one, see Figure 6.2. The operation is equivalent to a mirror reflection followed by a rotation through 180°.

If a system (a particle, or group of particles) is described by a wavefunction $\psi(x)$, then the parity operation will reverse the sign of the coordinates

$$\mathbf{P}\psi(x) = \psi(-x).$$

If the system is to remain invariant under the parity operation, the observable quantity which must not change is the probability density given, essentially, by the square of the wavefunction

$$\psi(x)\,\psi(x) = \psi(-x)\,\psi(-x).$$

Fig. 6.2. A parity transformation reverses the spatial coordinates of an event (*a*) or, equivalently, converts a left-handed coordinate system into a right-handed one (b).

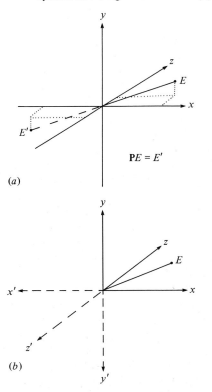

So we have

$$\psi(x) = \pm\psi(-x),$$

and so

$$\mathbf{P}\psi(x) = \pm\psi(x).$$

So if the system is to remain invariant under the parity operation, the system wavefunction may either remain unchanged $\mathbf{P}\psi = +\psi$, in which case we say the system is an *even* parity state; or the system wavefunction may change sign $\mathbf{P}\psi = -\psi$, in which case we say the system is an *odd* parity state.

If the forces governing the system respect parity, an even parity state cannot change into an odd one or vice versa. This helps us define the ways in which a system may evolve.

An example of this is the way in which light is emitted from atoms. Each state an electron can occupy has a definitive parity assignment, even or odd, which is determined by the magnitude of the orbital angular momentum of the electron about the nucleus and by the orientation of electron spin. As the electromagnetic force respects parity, transitions can be made only between states of the same parity. This limits the transitions possible and so prescribes the energies of the photons emitted. By observing the spectral lines emitted from atoms we can thus check the conservation of parity.

It is also necessary to consider the *intrinsic parity* of a single particle for which the operation of space inversion is not so obvious. This is illustrated by the decay of one particle into two. The final state of two particles with some well-defined motion with respect to one another can be examined under parity transformations and either even or odd parity assigned. If then the interaction responsible for causing the decay conserves parity, the initial one-article state must also be a state of well-defined parity. Thus a particle can be assigned some intrinsic parity, even $(+1)$ or odd (-1), which is multiplied together with the spatial parity to obtain the overall parity of the state.

Intrinsic parity has meaning only because particles can be created or destroyed. If the particles in a system were always the same, then the product of their intrinsic parities in any initial or final states would always be the same and so would be a meaningless quantity. In this hypothetical immutable world we should be free to assign any particle either even or odd parity with no reason. In the real world this arbitrariness allows us to define the intrinsic parity of certain particles (normally the nucleons are given even parity) and then the intrinsic parities of all other particles are established from experiment.

6.3.2 *Charge conjugation symmetry*

Another useful symmetry in particle physics is that of the interchange of particles with their antiparticles, denoted **C**. This symmetry means that if physical laws predict the behaviour of a set of particles, then they will predict exactly the same behaviour for the corresponding set of antiparticles. For example, a collision between an electron and a proton will look precisely the same as a collision between a positron and an antiproton, see Figure 6.3.

The symmetry applies also to the antiparticles of particles with no electric charge, such as the neutron. The interaction of a proton and a neutron is the same as that of an antiproton and an antineutron.

As with the parity operation, the wavefunction of a system may be even or odd under the action of the charge conjugation operation

$$\mathbf{C}\psi = \pm\psi.$$

An example of the use of this symmetry is provided by particle decay into photons by the electromagnetic force. A single photon is odd under **C** symmetry. Observing the decay of a particle into two photons then, determines the particle to be even under charge conjugation symmetry, as this is given simply by the product of the two photons' symmetries $(-1)^2$. We then know that the particle cannot

Fig. 6.3. Symmetry under the charge conjugation operation implies the equivalence of (*a*) particle and (*b*) antiparticle reactions.

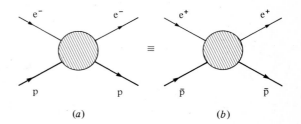

decay into an odd charge conjugate state, such as three photons, if **C** symmetry is to be preserved.

6.3.3 *Time reversal*

The last of the three discrete symmetries, denoted **T**, connects a process with that obtained by running backwards in time. Despite the rather intriguing name, the operation refers simply to that process obtained by reversing the directions of motion within the system. Symmetry under 'time reversal' implies that if any system can evolve from a given initial state to some final state, then it is possible to start from the final state and reenter the initial state by reversing the directions of motion of all the components of the system.

6.4 The CPT theorem

It is possible also to define product symmetries which can be obtained by operating two or more of these discrete symmetries simultaneously. For instance, a system of particles in a coordinate system can be subject to the operations of parity and charge conjugation simultaneously to reveal a system of antiparticles in the reverse-handed coordinate system. If the laws governing the system are invariant under the **CP** operation, then the two systems will behave in exactly the same way. Also it is possible to assign an even or odd symmetry under the combined **CP** symmetry to any state and so require that the system evolve to a state of the same symmetry.

There are no utterly fundamental reasons for supposing that the individual symmetries should be preserved by the various forces of nature (that there should be symmetry between a process and its mirror image, between a process and its antiprocess, and so on). But it seems a reasonable assumption and was taken for granted for many years. In fact, as we will soon see, the symmetries are *not* exact and there do exist phenomena which display slight asymmetries between process and mirror process, and process and antiprocess, see Chapter 12.

But there are very good reasons for supposing that the combined **CPT** symmetry is absolutely exact. So that, for any process, its mirror image, antiparticle and time-reversed process will look exactly as the original. This is the so-called **CPT** theorem which can be derived from only the most fundamental of assumptions, such as the causality

of physical events (cause must precede effect), the locality of interactions (instantaneous action at a distance is not possible) and the connection between the spin of particles and the statistics which govern their collective behaviour.

The consequences of the **CPT** theorem are that particles and their antiparticles should have exactly the same masses and lifetimes, and this has always been observed to be the case. Another consequence is that if any one individual (or pair) of the symmetries is broken, as mentioned, there must be a compensating asymmetry in the remaining operation(s) to cancel it and so ensure exact symmetry under **CPT**.

6.5 Dynamical symmetries

The symmetries of space and time give rise to universal conservation laws such as those of energy, momentum and angular momentum. As these laws must be respected by all processes, the Lagrangian of any system must be invariant under the groups of transformations through time, space, and angular rotations respectively.

But other conservation laws are also known to exist, such as the conservation of electrical charge. This can be represented by requiring the Lagran-

Fig. 6.4. A symbolic representation of the action of a dynamical symmetry.

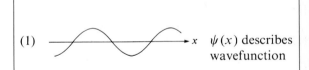

(1)　$\psi(x)$ describes wavefunction

(2) The Lagrangian describes interactions $L(\psi)$

(3)　$\psi(x)$　$\psi^*(x)$

A group of transformations **G** shifts the phase of the wavefunction
$$G\,\psi(x) = \psi^*(x)$$
$$G\,L(\psi) = L(\psi^*)$$

(4) Invariance requires $L(\psi) = L(\psi^*)$. This limits the possible functional form of L

gian to remain invariant under arbitary shifts in the phases of the charged particle wavefunctions appearing in the Lagrangian (see Figure 6.4).

We will learn that there are many other quantities which are conserved in interactions arising from the various forces of nature. This implies that the Lagrangians describing these interactions must be invariant under appropriate symmetry operations. We will see that demanding such invariances gives rise to physically significant predictions such as the existence of new particles and values for their electric charges, spins and other quantum numbers yet to be introduced.

6.6   Internal symmetries

The symmetry operations we have introduced so far are the fundamental ways of describing the conservation laws we observe to obtain in particle interactions. But symmetry can also help us categorise particles according to their intrinsic properties.

In addition to the familiar particles carrying only electrical charge, we will soon meet particles with wholly new quantum numbers such as 'strangeness', 'charm', and so on. The values of these quantum numbers carried by the particles allow them to be classified into fixed patterns or multiplets. We shall see this at work in Part 2.

Suffice at this stage to say that, in the microworld, symmetry does fulfil its traditional role of arranging disparate elements into regular patterns (just like the periodic table of elements).

6.7   Broken symmetries

Symmetries are sufficiently valuable that even broken ones can be useful. For many purposes a broken mirror is as good as a whole one! We have mentioned already how the individual reflection symmetries **P, C** and **T** might be broken in some classes of reaction (which turn out to be those governed by the weak nuclear force). But for other forces which do respect them, they are still a valuable guide for indicating which reactions are possible and which are forbidden.

Similarly, conservation laws and the internal symmetries on which they are based may not be exact. The first successful internal symmetry scheme for classifying the reactions of the strongly interacting particles was known from the start to be badly

broken but, nevertheless, it provided a valuable ordering effect on the variety of reactions observed.

Of particular interest is the case when the Lagrangian governing the dynamics due to some force or forces is not quite invariant under some group of gauge transformations, but only under a restricted group or when additional particles have been introduced. This indicates that the relatively more-complicated forces arising from the imperfect or restricted symmetries have their origin in a truly general symmetry (and its simpler forces) which may have obtained under different circumstances. This is the gist of the modern approach to the unified theory of the forces of nature in which approximate symmetries are a guide to the nature of forces in unfamiliar physical circumstances (e.g. just after the big bang).

7

Two mesons

7.1 Introduction

Modern particle physics can be thought of as starting with the advent of mesons. For these are not constituents of everyday matter, as are the protons and the electrons, but were first proposed to provide a description of nuclear forces. The subsequent discoveries of a bewildering variety of mesons heralded an unexpected richness in the structure of matter, which we are only now beginning to understand.

7.2 Yukawa's proposal

In attempting to describe the features of the strong nuclear force, physicists in the 1930s had to satisfy two basic requirements. Firstly, as the force acts in the same way on both protons and neutrons, it must be independent of electric charges and, secondly, as the force is felt only within the atomic nucleus, it must be of very short range. In 1935 the Japanese physicist H. Yukawa suggested that the nuclear force between protons is mediated by a massive particle, the pi-meson or pion, denoted by π, in contrast to the massless photon which mediates the infinite-range electromagnetic force. It is the mass of the mediating particle which ensures that the force it carries extends over only a finite range. This is indicated by Heisenberg's uncertainty principle which allows the violation of energy conservation for a brief period. If the proton emits a pion of finite mass, then energy conservation is violated by an amount equal to this mass energy. The time for which this situation can obtain places an upper limit on the distance which the pion can travel, and this distance is a guide to the maximum effective range of the force.

From the α-particle scattering experiments, we know that the effective range of the strong force is about 10^{-15} m, which gives a pion mass of about 300 times that of the electron, or about 150 MeV. To account for all the possible interactions between nucleons, the pions must come in three charge states. For instance, the proton may transform into a neutron by the emission of a positively charged pion or, equivalently, by the absorption of a negatively charged pion. But the proton may also remain unchanged during a nuclear reaction, which can be explained only by the existence of an uncharged pion. So the pion must exist in three charge states; positive, neutral and negative (π^+ π^0 π^-).

7.3 The muon

In 1937, five years after his discovery of the positron, Anderson observed in his cloud chamber yet another new particle originating from cosmic rays. The particle was found to exist in both positive- and negative-charge states with a mass some 200 times that of the electron, about 106 MeV. At first, the particle was thought to be Yukawa's pion and only gradually was this proved not to be the case. Most importantly, the new mesons seemed very reluctant to interact with atomic nuclei, as indicated by the fact that they are able to penetrate the earth's atmosphere to reach the cloud chamber at ground level. For particles which were expected to be carrying the strong nuclear force such behaviour was unlikely. Also, there was no sign of the neutral meson. Theorists eventually accepted that this new particle was not the pion, instead it was named the *mu-meson*, or *muon*, and is denoted by μ.

The muon was a baffling discovery as it seemed to have no purpose in the scheme of things. It behaves exactly like a heavy electron and it decays into an electron in 2×10^{-6} s; and so is not found in ordinary matter. Although we shall see later how the muon can fit into a second generation of heavy elementary particles, the reason for this repetition is still by no means obvious.

7.4 The real pion

If Yukawa's pion is to interact strongly with atomic nuclei, it is unreasonable to expect it to penetrate the entire atmosphere without being absorbed. So experiments at ground level are unlikely to detect it. In 1947 C. Powell, C. Lattes and G. Occhialini from Bristol University took photographic plates to a mountain top to reduce the decay distance that pions created at the top of the atmosphere had to traverse before being detected. They found Yukawa's pion, which quickly decays into a muon, which itself then decays, Figure 7.1. The mass of these charged pions $\pi^\pm$ was determined to be 273 times the mass of the electron (140 MeV), very close to Yukawa's original estimate. Since the initial discovery of the charged pions, it has been established that decay into a muon and a neutrino is their main decay mode with a lifetime of about 2.6×10^{-8} s. Other decay modes do exist but are thousands of times less likely.

The uncharged pion π^0 was eventually discovered in accelerator experiments in 1950. The delay was due to the fact that uncharged particles leave no obvious trace in most particle detectors and so cannot be observed directly. The most likely decay mode of the π^0 is into two photons which also leave no tracks. Only by observing the electron–positron pairs created by the photons can the existence of the π^0 be inferred, see Figure 7.2. The

mass of the π^0 was found to be slightly less than that of its charged counterparts at 264 times the mass of the electron, but its lifetime is much shorter at 0.8×10^{-16} s. The reason for this large difference in lifetimes is that the π^0 decays by the action of the electromagnetic force, as indicated by the presence of the two photons, whereas the charged pions decay by the weak force, as indicated by the presence of the neutrino.

In 1953 the pions were established spin 0 by comparing the relative magnitudes of the cross-sections of the reactions:

$$p+p\rightarrow d+\pi^+,$$
$$\pi^+ +d\rightarrow p+p.$$

The relative magnitude of the two can depend only on the spins of the particles in the collision, and knowing the spins of the protons (p) and the deuteron (d) determines that of the pion (π^+). Such reactions also establish the intrinsic parity of the pion (relative to the nucleons). This is found to be odd (-1).

7.5 Terminology

At this point it is worth both introducing some of the generic names which are used for these particles and defining their essential features. A few of the most often-used are:

nucleons: neutrons and protons;
hadrons: all particles affected by the strong nuclear force;

Fig. 7.1. The pion decays to a muon, which then decays to an electron. Neutrinos are emitted to ensure conservation of energy and angular momentum.

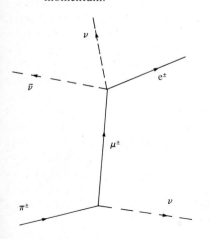

Fig. 7.2. The decay of the neutral pion.

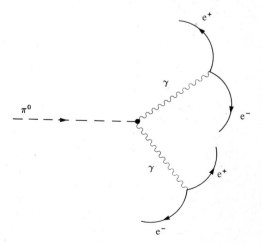

baryons: hadrons which are fermions (half-integral spin particles) such as the nucleons;

mesons: hadrons which are bosons (integral spin particles) such as the pion;

leptons: all particles not affected by the strong nuclear force, such as the electron and the muon.

Fig. 7.3. The analogy between particle spin in real space and particle isospin in imaginary space.

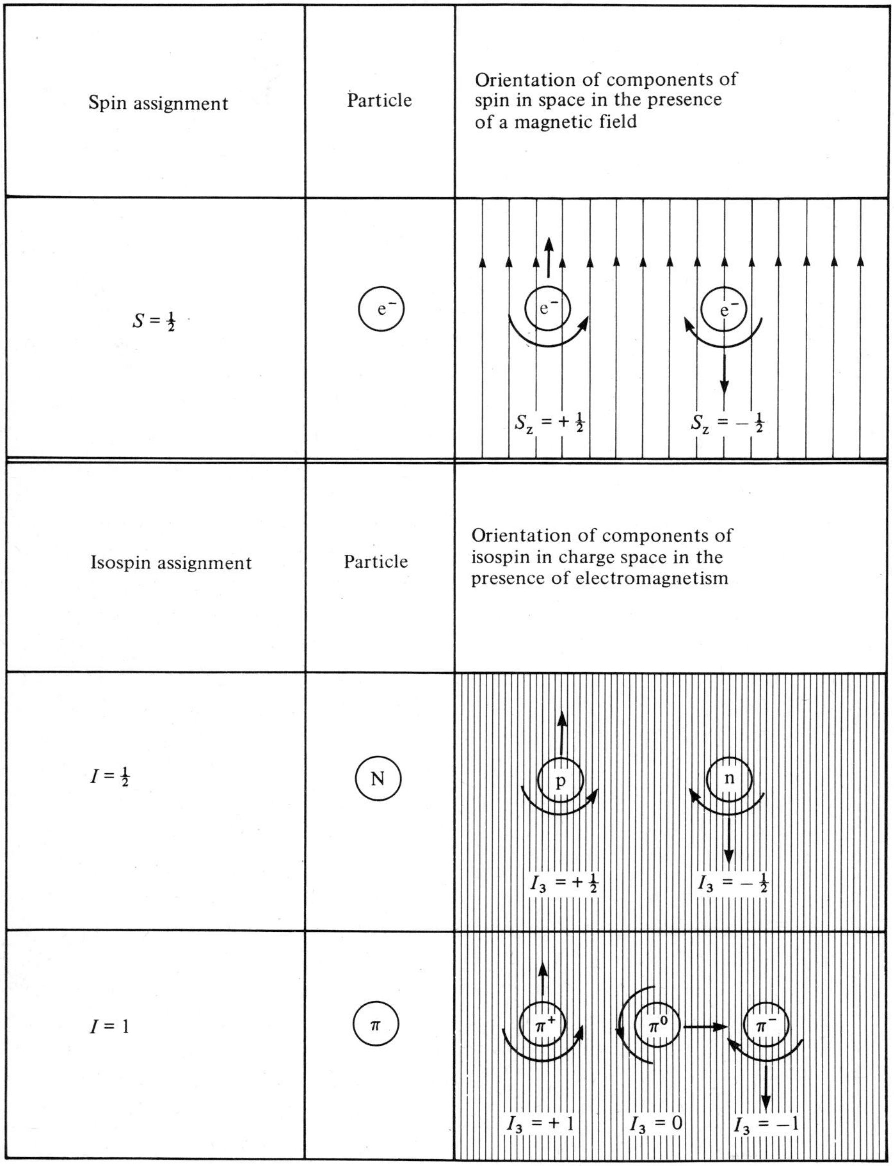

Particles which are baryons are assigned a baryon number B which takes the value $B=1$ for the nucleons, $B=-1$ for the antinucleons and $B=0$ for all mesons and leptons. In all particle reactions, the total baryon number is found to be conserved (i.e. the total of the baryon numbers of all the ingoing particles must equal that of all outgoing particles). Similarly, particles which are leptons are assigned a lepton number which is also conserved in particle reactions. This is explained further in Chapter 14.

Spin assignment	Particle	Orientation of components of spin in space in the presence of a magnetic field
$S=\frac{1}{2}$	e^-	$S_z=+\frac{1}{2}$ $S_z=-\frac{1}{2}$
Isospin assignment	Particle	Orientation of components of isospin in charge space in the presence of electromagnetism
$I=\frac{1}{2}$	N	$I_3=+\frac{1}{2}$ $I_3=-\frac{1}{2}$
$I=1$	π	$I_3=+1$ $I_3=0$ $I_3=-1$

7.6 Isotopic spin

We have met, so far, two sorts of particles which differ only slightly in their masses but which have different electric charges: the nucleon (the proton and the neutron), and the pions. The strong nuclear force seems to ignore totally the effects of electric charge and influences all nucleons in the same way and all pions in the same way. As far as the strong force is concerned, there is only one nucleon and only one pion. In 1932 Heisenberg described this mathematically by introducing the concept of *isotopic spin* or *isospin*. This concept is the prototype of both the elementary particle classification schemes and the modern dynamical theories of the fundamental forces, so it merits some attention.

Just as the two different orientations in real space of the 'third components' of the spin of the electron (see Chapter 3) provide two distinct states in which the electron can exist (in the presence of a magnetic field), Heisenberg proposed that different orientations in an abstract charge space of the third components of an imaginary isotopic spin would be a mathematically convenient way of representing the charge states within a family of particles (in the presence of electromagnetism) (see Figure 7.3). Similarly, just as the different components of electron spin are separated in energy by a magnetic field (causing the fine structure in spectral lines), so too the different components of isotopic spin in a particle family are separated in mass by the effects of the electromagnetic force (causing the slight mass differences between the proton and the neutron, and between charged and neutral pions).

The electric charges of the hadrons, Q, are related to their isospin assignments by the simple formula,

$$Q = e\left(I_3 + \frac{B}{2}\right).$$

So for the pions which have zero baryon number ($B=0$), the charges are simply the units of the electronic charge corresponding to the three 'third' components of spin I_3 (1, 0, −1). For the nucleon which has unit baryon number ($B=1$), the two isospin states with third component $+\frac{1}{2}$ and $-\frac{1}{2}$ become the positive and neutral charge states respectively.

8

Strange particles

8.1 Introduction

In 1947, the American physicists G. D. Rochester and C. C. Butler observed more new particles, about a thousand times more massive than the electron, in cloud chamber photographs of cosmic rays. As these particles were often associated with V-shaped tracks, they were at first called V particles, see Figure 8.1. Their origin and purpose were an entire mystery. Remembering that this same year saw the discovery of the real pion and the subsequent redundancy of the muon, it is fair to think of it as the beginning of the baroque era of particle physics, in which an increasing number of particles were discovered, seemingly with no other purpose than to decorate cloud chambers. For the following six years, the V particles were observed in cosmic ray experiments and two kinds become apparent. There are those whose decay products always include a proton and are called *hyperons*, and there are those whose decay products consist only of mesons and are called *K mesons*, or *kaons*.

The hyperons and kaons soon became known as the strange particles because of their anomalous behaviour. They were observed frequently enough to indicate production by the strong nuclear force, say, between two protons, or a pion and a proton, and so we would expect a decay time typical of a strong nuclear process (i.e. about 10^{-23} s). But, from the length of their tracks in the photographs, it was possible to estimate their average lifetimes at about 10^{-10} s, the timescale typical of weak interaction processes. This behaviour seemed to contradict the microscopic reversibility of reactions and required explanation.

8.2   Associated production

The first step in search of this explanation was provided in 1952 by the American physicist A. Pais. He suggested that the strange particles could not be produced singly by the strong interaction, but only in pairs. This was confirmed in experiments at the Brookhaven accelerator in 1953 when strange particles were man-made for the first time. The strange particles always emerged in pairs in reactions such as:

$$\pi^- + p \rightarrow \Lambda^0 + K^0,$$

where Λ^0 denotes the hyperon and K^0 denotes a neutral kaon.

In the same year Gell-Mann and Nishijima explained this mechanism of associated production by proposing the introduction of a new conservation law, that of strangeness, which applies only to the strong interaction. Each particle is assigned a quantum number of strangeness, in addition to its quantum numbers of spin, intrinsic parity and isospin. Then, in any strong interaction, the total

strangeness of all the particles before and after the reaction must be the same.

Associated production can now be explained by assigning a positive strangeness to one of the strange particles produced and a negative strangeness to the other so that the total strangeness of the final state is zero, the same as that of the non-strange initial state:

$$\pi^- + p \rightarrow \Lambda^0 + K^0 \qquad \text{Strangeness}$$
$$(0) + (0) \rightarrow (-1) + (+1) \qquad \text{assignments.}$$

The decay of strange particles into non-strange particles cannot proceed by the strong interaction which must, by definition, conserve strangeness. Instead, such decays proceed by the weak interaction, which need not, and which allows the strange particles a comparatively long life:

$$\Lambda^0 \xrightarrow[|\Delta S|=1]{} \pi^- + p \ (\tau \approx 10^{-10} \text{ s}).$$

The strangeness of the strongly interacting hadrons is defined by:

$$Q = e\left(I_3 + \frac{B+S}{2}\right).$$

When $S = 0$, we recover the equation of Section 7.6 which relates the charge to the third component of isospin for pions and nucleons and other non-strange hadrons.

8.3   The kaons

There are two charged strange mesons K^+ and K^- which each have a mass of 494 MeV, and a neutral one K^0 of mass 498 MeV. This makes the K mesons about three times more massive than the pions. But, like the pions, the kaons were found to be spin 0 and to have odd intrinsic parity. They are thus in some sense close relations of the pions. However, they have a very different multiplet structure. Let us recall that the three charge states of the pion (π^+, π^0, π^-) are the different I_3 states of the same $I = 1$ pion, and that the uncharged pion is its own antiparticle. This is not the case with the kaons because of complications due to the strangeness quantum number. If we assign to the neutral kaon K^0 a value of strangeness $S = 1$, then from the formula in Section 8.2, the value of total isospin $I = 1$ is ruled out and the kaons cannot form any isospin triplet like the pions. Instead, the kaons are

Fig. 8.1. (*a*) A neutral V^0 particle decays into pions. (*b*) A charged V^+ decays into a muon and a neutrino.

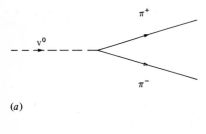

(a)

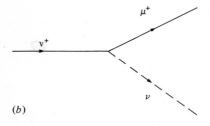

(b)

grouped into isospin doublets as shown in Figure 8.2. From this we can see that the uncharged kaon must come in two versions with opposite strangeness if the scheme is to work. So although the K^- is the antiparticle of the K^+, the K^0 is not its own antiparticle, which must have different strangeness:

$$\overline{K^+} \equiv K^- \text{ but } \overline{K^0} \neq K^0$$

Because the K^0 is different from the $\overline{K^0}$ only by the value of its strangeness, it might somehow be able to exhibit some effects directly attributable to strangeness. After all, so far we have merely categorised observed particle decay patterns by awarding the particles different values of a hypothetical quantum number. If we could observe some experimental effect due to strangeness, then we might be more convinced of its physical reality. This thought occurred at the time to Fermi who challenged Gell-Mann to prove the worth of his strangeness by demonstrating some difference between the K^0 and the $\overline{K^0}$. This led to some very important work, as we shall see in Chapter 15.

We can very neatly summarise our knowledge of the mesons discussed so far by plotting them on a graph where the axes show their assignments of isospin and strangeness, Figure 8.3. These graphs are known as multiplets for particles of the same spin and intrinsic parity and we will see how they form the basis of the elementary particle classification scheme in Chapter 11.

8.4 The hyperons

The hyperons are the strange particles which eventually decay into a proton and which, like the proton, have spin $\frac{1}{2}$ and are baryons with baryon number 1. The lambda hyperon Λ^0 is the least

massive at 1115 MeV and has isospin zero (it exists only as a neutral particle). The sigma hyperon Σ has a mass of 1190 MeV and has isospin 1 and so exists in three different charge states ($\Sigma^+, \Sigma^0, \Sigma^-$). Finally, the xi hyperon Ξ, known also as the cascade particle, has mass 1320 MeV and isospin $\frac{1}{2}$ and has strangeness -2. To decay into non-strange particles, it therefore needs to undergo two weak interactions, as the weak force can only change strangeness by one unit at a time.

$$\Xi^0 \rightarrow \Lambda^0 + \Pi^0 \qquad |\Delta S| = 1$$
$$ \hookrightarrow \pi^- + p \quad |\Delta S| = 1.$$

For the hyperons, we often prefer to use hypercharge Y as the distinguishing quantum number, which is the sum of baryon number and strangeness,

$$Y = B + S.$$

Those Λ, Σ and Ξ hyperons of spin $\frac{1}{2}$ which have been mentioned form only the basic set of those which exist. There are very many more massive hyperons which have spins $\frac{3}{2}$, $\frac{5}{2}$ or even $\frac{7}{2}$. These resonances are short lived and generally decay quickly into one of the hyperons in the basic set by the strong interaction (conserving strangeness, or hypercharge) before these eventually decay by the weak interaction back into non-strange baryons.

Fig. 8.3. A multiplet of pions and K mesons arranged according to value of strangeness and third component of isospin.

Fig. 8.2. Isospin doublets of the K mesons.

$$S = 1, \; I = \tfrac{1}{2} \quad \begin{cases} I_3 = +\tfrac{1}{2} & K^+ \\ I_3 = -\tfrac{1}{2} & K^0 \end{cases}$$

$$S = -1, \; I = \tfrac{1}{2} \quad \begin{cases} I_3 = +\tfrac{1}{2} & \overline{K^0} \\ I_3 = -\tfrac{1}{2} & K^- \end{cases}$$

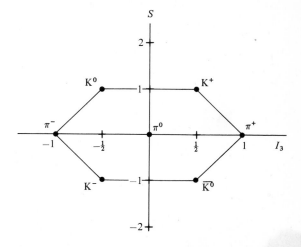

8.5 Summary

In Figure 8.4, all the particles we have mentioned so far are plotted according to their masses and are categorised according to their generic names. The origin of the names is clear from the diagram, the leptons are the lightweights, the mesons are the middleweights and the baryons are the heavyweights. We also show the applicability of the fundamental forces to the various categories of particles. We may think it more than just coincidence that the strongly interacting hadrons are the most massive category if we believe that the mass of the particles somehow arises from the interactions they experience.

We now know that mass alone is not a reliable way to categorise the species. Recent experiments have found both leptons and mesons more massive than the baryons. Nowadays the classifications are taken to refer to the interactions experienced by the various classes, which is taken to be a more fundamental attribute than mass.

Fig. 8.4. The basic set of elementary particles known by the early 1950s.

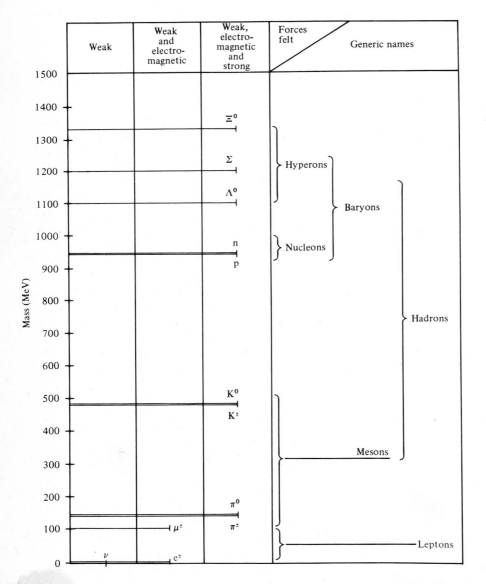

Part 2
Strong interaction physics

Resonance particles

9.1 Introduction

Most of the particles which we have discussed up to this point have lifetimes sufficient for them to leave observable tracks in bubble chambers or other detectors, say greater than about 10^{-12} s. But there is no reason for us to demand that anything we call a particle should necessarily have this property. It may be, for instance, that some particles exist only for a much shorter time before decaying into others. In this case we should not expect to detect them directly, but to have to infer their existence from the indirect evidence of their decay products. These transient particles are called *resonance particles* and many have been discovered with widely varying properties. It was the attempts to categorise the large number of resonances which first led to an appreciation of the need for a more fundamental pattern of order, which in turn led to the idea of quarks.

9.2 Resonance particle experiments

Resonance particles can be produced in two different types of experiment: resonance formation and resonance production experiments. In the formation experiments, two colliding particles come together to form a single resonance which acts as an intermediate state between the original colliding particles and the final outgoing products of the collision. The presence of the resonance is indicated when the cross-section for the collision varies dramatically over a small range of collision energy centred on the mass of the resonance, see Figure 9.1. The value of the energy range corresponding to one-half of the height of the resonance peak is referred to as the width of the resonance and this is a measure of the uncertainty in the mass of the particle.

Only if a particle is perfectly stable can it be thought of as having a uniquely defined mass, as time is needed for the act of measurement defining the mass. For an unstable particle there will always be uncertainty in the value of its mass, given by Heisenberg's uncertainty principle

$$\Delta E \Delta t \geqslant \hbar.$$

From this we can see that the narrower the width ΔE of the resonance, the larger will be the uncertainty in the lifetime, thereby implying a longer-lived particle. Conversely, if the resonance is broad, this implies a short lifetime. Typical widths for hadronic resonances, such as the N* resonances in pion–proton scattering are a few hundred MeV, which correspond to lifetimes of about 10^{-23} s. This makes them the most transient phenomena studied in the natural world.

In resonance production experiments, the presence of a resonance is inferred when it is found that the outgoing particles prefer to emerge with a particular value of combined mass. Finding the resonances in this fashion is more difficult because it is first necessary to look at all the possible combinations of outgoing particles which might have arisen from the resonance, and then to plot the combined masses of the combinations to see if any preferred values exist, see Figure 9.2.

One advantage of the production method is that it does not require us to study only the resonances which can be formed by the ingoing particles. In high-energy collisions, any number of new and interesting particles may emerge and it is possible to see if they have originated from some previously unknown resonance. In this fashion we

Fig. 9.1. An example of resonance formation. A large increase in the pion–proton cross-section plotted against the pion beam momentum $P_{\pi\text{beam}}$ signals the formation of a resonance particle.

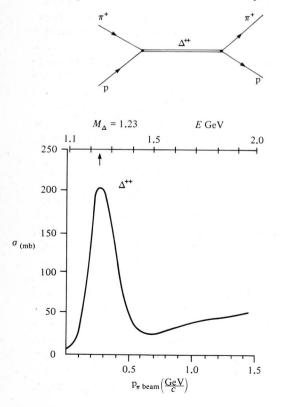

Fig. 9.2. An example of resonance production.

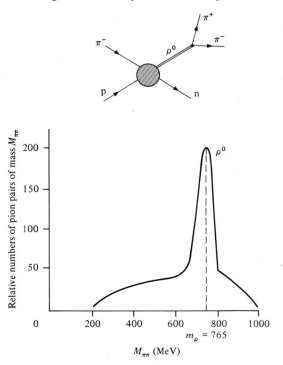

can study resonances made from πs and Ks only, even though we cannot arrange collisions between only πs and Ks as the ingoing particles. These methods have allowed physicists to build up a picture of literally hundreds of resonances, all of which may be legitimately regarded as just as elementary as the neutron or the pion.

Over the years, interesting regularities in the resonance spectrum became apparent. Often a particular set of quantum numbers for isospin and strangeness, such as for the pion ($I=1$, $S=0$) and the kaons ($I=\frac{1}{2}$, $S=1$) are duplicated by particles with higher masses and spins. These higher-mass versions of the quantum numbers generally decay very quickly back down to the least-massive particle with those particular numbers, by the strong interaction, and this least-massive version can then itself decay more slowly by the weak force, violating quantum-number conservation as it does so, see Table 9.1.

Table 9.1. *Two mass series of meson resonances*

Meson symbol	I	S	Mass (MeV)	Spin	Decay	Force acting
π	1	0	140	0	$\mu\nu$	weak
ρ	1	0	765	1	$\pi\pi$	strong
A_2	1	0	1300	2	$\rho\pi$	strong
g	1	0	1690	3	4π	strong
K	$\frac{1}{2}$	1	493	0	$\mu\nu$	weak
K*	$\frac{1}{2}$	1	892	1	$K\pi$	strong
K**	$\frac{1}{2}$	1	1420	2	$K\pi$	strong

10

SU(3) and quarks

10.1 Introduction

By the early 1960s it became clear that many hundreds of so-called 'elementary' resonance particles exist, each with well-defined values of the various quantum numbers such as spin, isospin, strangeness and baryon number and with widths which are generally seen to increase (or lifetimes which are generally seen to decrease) as their masses become larger. At that time, the most urgent task for physicists was to discover the correct classification scheme for the particles, which would do for the elementary particles what Mendeleeff's periodic table had done for the variety of elements known in the nineteenth century. A closely related problem was whether or not it was sensible to regard such a plethora of different particles as truly elementary. Most of the resonance particles are very massive compared, say, to the electron, occupy a finite region of space with a radius of about 10^{-15} m and many have high values of spin and the internal quantum numbers. All these factors argue in favour of the existence of more-fundamental constituents combining in a variety of ways to make up the known hadrons, just as a few fundamental atomic constituents (electrons, protons and neutrons) can combine to make up the variety of elements. But, historically, it was not possible to pass directly to the analysis of these fundamental constituents, which at the time were extremely speculative. Initially, it was necessary to classify the bewildering variety of hadrons according to some symmetry scheme from which clues to the nature of the constituents could be derived.

10.2 Internal symmetry

Such a classification scheme is provided by an internal symmetry group, as described in Chapter 6. The starting point of this symmetry group is the charge independence of the strong nuclear forces, as expressed in Chapter 7 by the concept of isospin. By regarding the neutron and the proton as the isospin down and isospin up components of a single nucleon, the strong interaction's indifference to 'neutron-ness' and 'proton-ness' can be expressed as the invariance of strong interactions to rotations in the imaginary isospin space. The group of rotations which achieve these rotations is the *Special Unitary* group of transformations of dimension 2 called $SU(2)$ which act on the 2-dimensional space defined by the proton and the neutron, redefining the proton and neutron as a mixture of the original two,

$$\mathbf{G}^{SU(2)}\begin{pmatrix} p \\ n \end{pmatrix} \rightarrow \begin{pmatrix} p' \\ n' \end{pmatrix}.$$

Of course, the same must also be true of the pions, which form a 3-dimensional space ($\pi^+ \pi^0 \pi^-$), and the Δ baryons (Δ^{++}, Δ^+, Δ^0, Δ^-), which form a 4-dimensional space. These are referred to as the 2-, 3- or 4-dimensional representations of $SU(2)$.

When conservation of strangeness is added to that of isospin as a property of the strong interaction, it is clear that the strongly interacting particles are governed by a bigger symmetry group. Although seemingly obvious, it took a great deal of work to show that $SU(3)$ is the appropriate group. The transformations of the $SU(3)$ group generate many dimensional representations (multiplets), **3**, **6**, **8**, **10**, **27**, etc., each of which is a well-defined quantum-number pattern. It was a triumph for the originators of the scheme to find that some of these exactly fitted the quantum-number structure of the observed hadrons, see Figure 10.1. The identification of the correct symmetry group for the strong interactions, and the assignment of hadrons to the multiplets led to the prediction in 1962 of a new hadron necessary to complete the spin-$\frac{3}{2}$ baryon decuplet **10**. This is the famous Ω^- particle with strangeness assignment -3. Its spectacular discovery in 1963 in bubble chamber photographs at

Brookhaven convinced a previously sceptical world of the validity of $SU(3)$.

Having found the correct symmetry group, a major problem remained. It was necessary to explain why the mesons filled some multiplets, the baryons fitted others, but other multiplets had no particles. In particular it seemed odd that the

Fig. 10.1. $SU(3)$ representations provide the quantum-number patterns for the elementary particles.

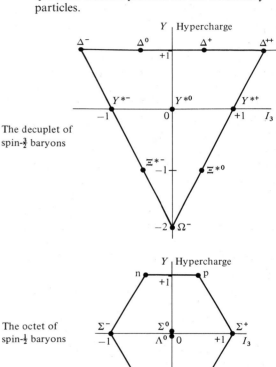

The decuplet of spin-$\frac{3}{2}$ baryons

The octet of spin-$\frac{1}{2}$ baryons

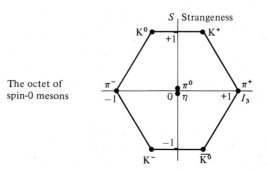

The octet of spin-0 mesons

fundamental 3-dimensional representation should remain unfilled (i.e. the most basic representation of the $SU(3)$ group). In an unsuccessful symmetry scheme prior to that of Gell-Mann's, the proton, neutron and hyperon were assigned to this triplet, but the logical consequences of such an assignment were incompatible with the experimental evidence.

10.3 Quarks

In 1964, Gell-Mann & Zweig pointed out that the representations of $SU(3)$ which were occupied by particles could be chosen from amongst all those mathematically possible by assuming them to be generated by just two combinations of the fundamental representation. Gell-Mann called the entities in the fundamental representation *quarks*, this being the rather idiosyncratic use of a German word meaning curds or slop (abstracted for the purpose from the novel *Finnegan's Wake* by James Joyce). The three varieties of quark, or flavours as they are now called, have since come to be known as the up, down and strange quarks, the up and down labels referring to the orientation of the quarks' isospin. The combinations of quarks which give the occupied representations of $SU(3)$ are a quark–antiquark pair for the meson multiplets and three quarks for the baryon multiplets. This is expressed mathematically by combining the representations of the group

$$\mathbf{q} \otimes \mathbf{q} \otimes \mathbf{q} \equiv \mathbf{3} \otimes \mathbf{3} \otimes \mathbf{3} \rightarrow \mathbf{1} \oplus \mathbf{8} \oplus \mathbf{8} \oplus \mathbf{10},$$

$$\mathbf{q} \otimes \bar{\mathbf{q}} \quad \equiv \mathbf{3} \otimes \mathbf{3}^* \quad \rightarrow \mathbf{1} \oplus \mathbf{8}.$$

The quark constituents of the baryon decuplet and of the meson octet are illustrated in Figure 10.2.

One significant consequence of this scheme is that if three quarks are to make up each baryon with a baryon number 1, then the quarks themselves must have baryon number $\frac{1}{3}$. From the formula relating charge to isospin and baryon number, this means that they must also have fractional electronic charge. Also, to ensure that the baryons generated are fermions and the mesons, bosons, it is necessary to assign the quarks spin $\frac{1}{2}$. A summary of their properties is shown in Table 10.1.

The quarks were referred to earlier as entities rather than particles for good reason. It is not necessary to assume their existence as observable

particles to enjoy the successes of the $SU(3)$ flavour scheme. They may be thought of as the mathematical elements only for such a scheme, devoid of physical reality. This was a fortunate escape clause at the beginning of the quarks' career because their fractional electronic charges and the failure to observe them in experiments encouraged scepticism in the naturally conservative world of physics. As

Fig. 10.2. The quark content of the $SU(3)$ representations. (qqq)′ signifies the summation over the cyclic permutations of the quarks.

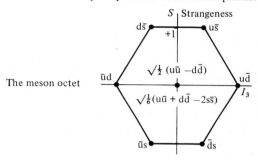

The meson octet

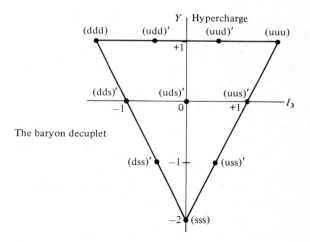

The baryon decuplet

Table 10.1. *The quantum number assignments of the early quarks*

Quark	q	Spin	Charge	I	I_3	S	B
Up	u	$\frac{1}{2}$	$+\frac{2}{3}$	$\frac{1}{2}$	$+\frac{1}{2}$	0	$\frac{1}{3}$
Down	d	$\frac{1}{2}$	$-\frac{1}{3}$	$\frac{1}{2}$	$-\frac{1}{2}$	0	$\frac{1}{3}$
Strange	s	$\frac{1}{2}$	$-\frac{1}{3}$	0	0	-1	$\frac{1}{3}$

we shall see, indirect evidence for the physical reality of quarks is now very convincing – despite the fact that they have never been seen directly in isolation. But this evidence has mounted only rather slowly since 1968 with the beginning of the 'deep inelastic' experiments at the Stanford Linear Accelerator Center (SLAC) in California. Prior to this, most physicists preferred to reserve judgement on the reality of quarks, content to rely on the mathematics of $SU(3)$ only.

Because of this historical background of doubt, conclusions which rely on the mathematics of group theory are put together as the $SU(3)$ scheme of the elementary particles, and conclusions which rely on the physical reality of quarks are referred to as the 'quark model'. The mathematics of $SU(3)$, in addition to generating the multiplet structure of the observed particles, can also provide simple predictions of relationships between the masses of the particles in an $SU(3)$ multiplet. If the $SU(3)$ symmetry were perfect, then all the particles in the same $SU(3)$ multiplet would have to have the same mass. This is obviously not true and so we know that $SU(3)$ cannot be a perfect symmetry: it is broken. But by making assumptions about just how the symmetry fails, it is possible to derive mass formulae which seem to hold good.

$$\tfrac{1}{2}(m_N + m_\Xi) = \tfrac{1}{4}(3m_{\Lambda^0} + m_\Sigma) \quad \text{baryons,}$$
$$m_K^2 = \tfrac{1}{4}(3m_\eta^2 + m_\pi^2) \quad \text{mesons.}$$

In the quark model, the effects of symmetry breakdown can be described by saying that the strange quark has a larger mass than either of the equal mass up and down quarks. Also, in the quark model, it is possible to assume the existence of forces holding the quarks together and then to generate the spectrum of elementary particle masses for particles of the same quantum number and different spins. The predictions are found to match the experimentally measured masses really rather well, better than the approximations of the model would seem to justify in fact. But the simple quark model is unable to explain the outstanding problems surrounding the quarks. Why are they not seen? Why do they seem to form only in certain combinations? What are the nature of the forces which they experience? These questions, as we will see, await the advent of a theory of quarks on a par with the QED theory of electrons.

11

Hadron dynamics

11.1 Introduction

The unresolved problems of the quark model led the majority of physicists in the 1960s to restrict their attention to the dynamics of the observed hadrons only. Although it was generally realised that no fundamental theory of the strong interaction was likely to result, this work did lend some order to the bewildering variety of effects seen in hadron collisons, such as those between pions and protons (πp), kaons and protons (Kp) and protons and protons (pp) which constituted the majority of the experimental programmes for the 1960s. This investigation into hadron dynamics performs the complementary task to the $SU(3)$ internal symmetry scheme which orders the static properties of the hadronic resonance spectrum.

11.2 Regge trajectories

In simple two-particle scattering, Regge theory describes how the angular momentum of the two-particle system varies with the energy of the collision. So, in the resonance region, Regge theory tells us how the spin of the resonances varies with their mass. If we know the form of the forces acting between the particles, we can predict the spectrum of resonances on a graph of spin against mass. These are known as *Regge trajectories*, see Figure 11.1.

Regge theory was first used for the low-energy scattering of particles in a known potential, such as in the electric field in an atom. Strictly speaking, the theory can be justified only in this area. It came as a pleasant surprise to physicists to find that the resonances observed in higher-energy collisions lay on particularly simple Regge trajectories. They are all straight lines. This provides a classification scheme which relates the spins and the masses of all resonances with the same internal quantum numbers, in contrast to the $SU(3)$ scheme which relates particles of different quantum numbers with the same values for spin (and mass in perfect $SU(3)$). Some of the Regge trajectories for mesons and baryons are shown in Figure 11.2.

11.3 Hadron collisions

The usefulness of Regge theory does not stop simply at providing another classification scheme for the resonances, it can describe hadron collisions also. Looking back to Chapter 7, we remember that the earliest picture of the strong nuclear force was that of it being carried between two colliding hadrons (say protons) by a pion, in analogy with the photon exchange mechanism of an electromagnetic process. Unfortunately, this simple picture has severe limitations. Firstly, the strength of the nuc-

lear forces means that there is no justification for ignoring two-, three- or more pion exchange processes; in principle they may be considerably more important than simple pion exchange. Secondly, the calculations which can be performed for single-pion exchange give only limited agreement with experimental results.

And finally, one pion exchange cannot alone suffice: many different mesons are known and it is unlikely that these have no part to play in hadron dynamics. Also, there are some types of collision (pion–proton scattering, for example), where pion exchange is actually forbidden by conservation of parity or some other quantum number. This indicates that some other mechanism must be at work.

An improved description of high-energy hadron collisions is obtained when it is not just one pion (or any other meson) which is taken as being exchanged by the colliding particles, but the entire

Fig. 11.1. Particles approach with given energy and angular momentum to form a resonance of given mass and spin. The spectrum of resonances lies on a Regge trajectory relating mass and spin.

Fig. 11.2. Regge trajectories for mesons and baryons.

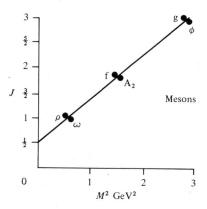

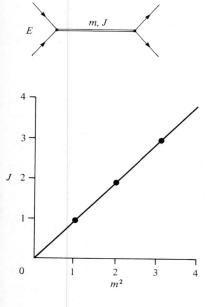

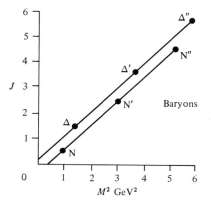

Fig. 11.3. The exchange of a reggeon is equivalent
to the exchange of many different spin
resonances.

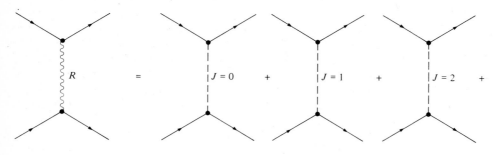

Fig. 11.4. Examples of the behaviour of
cross-section data obtained in high-energy hadron
collisions.

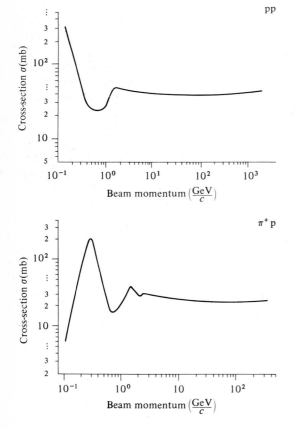

Fig. 11.5. Multiplicity data in hadron collisions.

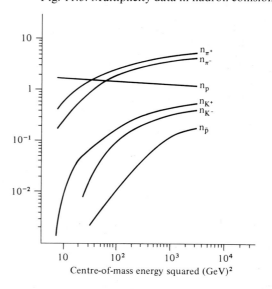

contents of one or more Regge trajectories. This is often referred to as *reggeon* exchange. A single reggeon exchange will therefore represent the exchange of all those resonances with differing masses and spins with an otherwise identical set of internal quantum numbers, see Figure 11.3.

The calculations of the reggeon exchange process are far less certain than the Feynman rules of QED, but they do a similar job. In particular they predict that the total cross-sections for hadron collisions are nearly constant over a very wide range of collision energies, just as is observed in experiments, see Figure 11.4.

For hadron collisions at very high energies, say at about 50 times the rest mass energy of the proton, the dominant feature of most events is the sheer number of particles which are produced. This number is called the multiplicity of the reaction, see Figure 11.5. Regge theory can be extended to describe these reactions by introducing the concept of multiple reggeon exchange. Formulae describing these processes can then be used as the basis for a reggeon field theory of hadron collisions in which it is the reggeons that play the same role as do the photons in QED.

This picture of hadron collisions is made consistent with the $SU(3)$ classification scheme of the hadrons by using quark diagrams to show the flow of internal quantum numbers in a collision process whose dynamics are described by Regge theory. Thus it is possible to represent electric charge or strangeness being carried by the reggeon from one external particle to another, see Figure 11.6. As mentioned previously, the use of quarks and quark line diagrams in no way requires the physical existence of the quarks – they may be thought of purely as convenient mnemonics for the $SU(3)$ symmetry of the observed hadrons. On the other hand, the progression to quark line diagrams is highly suggestive of a more fundamental role.

11.4 Summary

It was the plethora of resonance particles discovered mainly in hadron collisions which first gave rise to the two classification schemes of $SU(3)$ flavour symmetry and Regge trajectories. The $SU(3)$ scheme successfully relates particles of different quantum numbers and of the same spin and in so doing introduces elemental quantum-number entities which act as building blocks, certain combinations of which can account for all the observed hadrons. The entities may be pure abstraction or may indeed represent new fundamental particles called quarks. The Regge theory relates hadrons of different spins and masses but of the same quantum numbers. It also provides a mathematical description of high-energy hadron collisions, but with no fundamental justification. The quark model enters only as a quantum-number book-keeper for Regge theory if quarks do not exist. But if they do, then Regge theory is only a 'macroscopic' approximation to the real dynamics of the fundamental quarks.

Fig. 11.6. Quark line diagrams corresponding to reggeon diagrams.

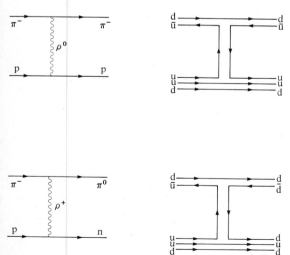

Part 3

Weak interaction physics I

The violation of parity

12.1 Introduction

The decay of the strange kaons led to a great deal of confusion in the early 1950s. Two decay modes in particular seemed so different that they were at one time thought to originate from two different parent particles, called the τ and θ mesons:

$$\tau^+ \to \pi^+ + \pi^+ + \pi^-,$$
$$\theta^+ \to \pi^+ + \pi^0.$$

However, detailed study of the two- and three-pion final states indicated that the τ and θ were indeed both manifestations of the same charged kaon. In both cases the mass was the same, and so too was the lifetime – about 10^{-8} s, a timescale which indicates it is the weak force that is responsible for the decays. The decays were thought to be incompatible because the parities of the two final states are different. If they originate from the same initial particle, they imply that parity is not conserved by the force responsible for the decays. This means that the force behaves differently in left-handed and right-handed coordinate systems: it can distinguish left from right, or image from mirror image.

Such a revolutionary conclusion was not seriously entertained until 1956, when T. D. Lee and C. N. Yang pointed out that, although evidence existed for the conservation of parity by the strong and electromagnetic forces, there was no evidence for its conservation by the weak force. Certainly, the τ–θ puzzle indicated that the weak force did not conserve parity, and Lee and Yang proposed that this was a general feature of all weak interactions.

12.2 β decay of cobalt

Within months of Lee and Yang's suggestion, experiments were performed to test for parity violation in other weak processes. The first and most famous was conducted on the β decay of cobalt by C. S. Wu and E. Ambler at the National Bureau of Standards in Washington. The point of the experiment was to observe some spatial asymmetry in the emission of β-decay electrons from the

cobalt, which could lead to a distinction between β decay and its mirror-image process. The process in question was the ordinary radioactive β decay of cobalt into nickel;

$$Co^{60} \rightarrow Ni^{60} + e^- + \bar{\nu}.$$

Firstly it was necessary to establish some direction in space of which the cobalt was aware, and with respect to which the emission of β-decay electrons could be measured. This was done by putting a magnetic field across a specimen of cobalt, cooled to a very low temperature. In this situation, the spin of the nuclei align predominantly along the direction of the magnetic field. By measuring the emission of the β-decay electrons along or against the orientation of nuclear spin (the orientation of the magnetic field), any asymmetry can be detected. It is possible to show that the direction of spin will not

Fig. 12.1. (*a*) If no asymmetry were detected in the emission of decay electrons, the real world and the mirror world would be indistinguishable. (*b*) If asymmetry were detected, this would result in a distinction between the two. This latter case is observed in experiments.

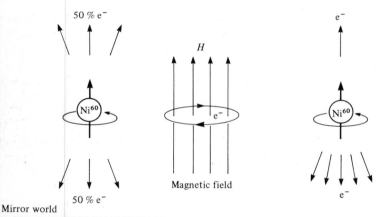

Mirror world

Real world

(*a*)   (*b*)

change under mirror reflection, nor will the direction of the magnetic field. But the direction in which the β-decay electrons are emitted will change under mirror reflection and so any asymmetry of electron emission measured with respect to the magnetic field direction will appear to be reversed in the mirror-image process, see Figure 12.1. Hence the process and its mirror image are distinguishable, and the weak force responsible for nuclear β decay can tell its right hand from its left.

The world of physics could scarcely have been more surprised when Wu and her colleagues duly observed the asymmetry which Lee and Yang's work had implied. Not since the discovery of the quantum nature of light had Nature seemed so contrary to common perception. The shock is reputed to have led one eminent physicist to accuse God of being 'weakly left-handed'.

Other experiments soon confirmed this parity-violating effect. One example is provided by the decay of the hyperon in the process,

$$\pi^- + p \rightarrow \Lambda^0 + K^0$$
$$ \hookrightarrow \pi^- + p.$$

It is possible to define a plane formed by the tracks

Fig. 12.2. How the basic Co^{60} experiment transforms under repeated **C** and **P** transformations.

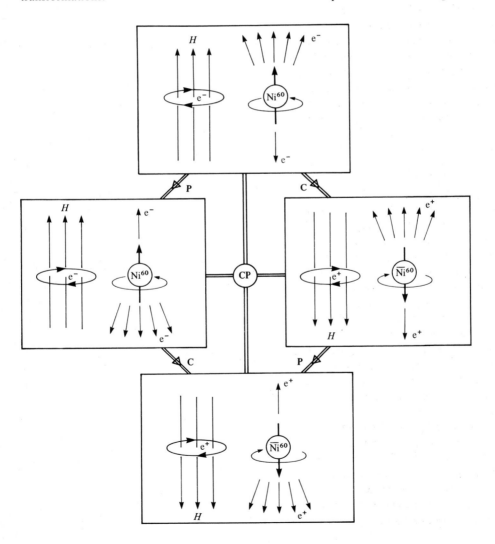

of the incoming π^- and the outgoing hyperon. If parity were conserved, equal numbers of outgoing π^- would emerge on either side of this plane. In 1957 an experiment yet again detected an asymmetry indicating parity violation.

12.3 Absolute-handedness and CP invariance

It is an interesting, if rather academic, question of philosophy to ask whether or not it is possible to distinguish absolutely between left and right, using the parity-violating effects of the weak force. The famous thought-test of such a distinction is to attempt to communicate our convention for left and right (or clockwise and anticlockwise) to an intelligent alien in a distant galaxy.

We might think of achieving this by instructing the alien to perform the Co^{60} experiment and telling him that our definition of anticlockwise is the direction required of the electron loop that provides the magnetic field, when viewed from the direction towards which most of the β-decay electrons are emitted. This would certainly suffice for aliens in our galaxy, but would not necessarily for aliens in more-distant parts of the universe. This is due to the possibility that distant aliens may be made of antimatter and may be conducting an anticobalt 60 experiment in which precisely the same procedures would lead to the opposite of our intended conclusions. The reason for this is that although the weak force violates parity, it also violates the symmetry of charge conjugation (matter–antimatter interchange) in such a way that the product symmetry of the two, denoted **CP**, is almost exactly conserved.

Starting from our original experiment in which the majority of β-decay electrons emerge in the direction of the field, we can see that the operation of space inversion will lead to an observable difference: the electrons are emitted against the direction of the field. However, if we then imagine the *additional* operation of charge conjugation we are led to an exact copy of the original process: the particles are emitted along the direction of the field, see Figure 12.2. So a real-matter alien looking at our original experiment from the direction in which most decay products are emitted will see a clockwise current loop, but the antimatter alien will see an anticlockwise current loop. Thus an alien observing that the emitted particles come out preferentially in the direction of the field would know that *either* his

conventions about left–right and particles–antiparticles were the same as ours *or* that they were both different.

Of course, we have not been able to test the validity of the **CP** conservation using anticobalt, although other experiments have been conducted to show that it is preserved to a high degree. But this is not the end of this particular story, as we will see in Chapter 15.

13

Fermi's theory of the weak interactions

13.1 Introduction

Prior to the early 1960s, just three different leptons were recognised. The electron, the muon and the neutrino (together with their antiparticles). The best place to study the weak interaction is in processes involving these leptons only. This ensures that there are no unwanted strong interaction effects spoiling the picture. Unfortunately, early opportunities to study purely leptonic reactions were limited, being restricted to the muon decay into an electron and neutrino. The most common weak interactions available for study are the radioactive β decay of nuclei and the decay of the pions and kaons (which are described generically as the weak decay of hadrons), and it was predominantly these reactions which formed the basis for the first description of the weak interactions formulated by Fermi in 1933.

13.2 Fermi's theory of β decay

The simplest manifestation of β decay is the decay of a free neutron into a proton, an electron and an antineutrino, see Figure 13.1.

$$n \rightarrow p + e^- + \bar{\nu}.$$

Fermi took this to be the prototype for the weak interactions, which he thus described as four fermions reacting at a single point. He expressed this mathematically by saying that, at a single point in space–time, the quantum-mechanical wavefunction of the neutron is transformed into that of the proton and that the wavefunction of the incoming neutrino (equivalent to the outgoing antineutrino we actually see) is transformed into that of the electron. So a description of this reaction is provided by multiplying the wavefunctions by unknown factors Γ which effect the transformations, and by another factor G_F called the Fermi coupling constant. This is the quantity which governs the intrinsic strength of the weak interactions, and so the rate of the decay. Thus the amplitude for β decay is given by,

$$M = G_F \, (\bar{\psi}_p \Gamma \psi_n) \, (\bar{\psi}_e \Gamma \psi_\nu).$$

The factors Γ contain the essence of the weak interaction effects which give rise to the transformation of the particles. The challenge was to discover the nature of these quantities (whether they are just numbers (scalars) or vectors, tensors, etc.). By examining the angles of emission between the outgoing products of β decay and their various energies, it is possible to narrow down the choice. This took many years: their nature was not confirmed until the parity-violating effect of the weak force was known.

In 1956, Lee and Yang proposed that the interaction factors Γ be a mixture of vector and axial vector quantities, to account for the parity-violating effects of the interactions.

A vector quantity has well-defined properties under a Lorentz transformation. For instance, it will change sign if rotated through 180° and will appear identical after rotation through 360°. An axial vector quantity will transform just like a vector under rotations, but will transform with the opposite sign to a vector under improper transformations such as parity. Thus, if the interaction is comprised of vector and axial vector components, it

Fig. 13.1. The β decay of a free neutron.

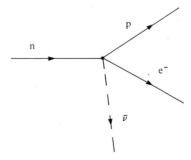

will look different after a parity transformation (the components might add together instead of cancelling), which is just what we need to describe the weak interactions. By inserting this form of interaction factor Γ into the amplitude M for β decay, it is possible to calculate the features of particle emission in free neutron decay.

13.3 The polarisation of β-decay electrons

We have already seen how parity violation manifests itself in β decay as an asymmetry in the direction of emission of electrons. But it also affects the spins of the emitted electrons. In the absence of parity violation, as many left-handed as right-handed electrons should be emitted, see Figure 13.2. But because of it, the electrons show a net preference to spin in one of the two possible ways. We define a polarisation P of the electrons to quantify this preference:

$$P = \frac{N_R - N_L}{N_R + N_L},$$

where N_L (N_R) is the number of left-spinning (right-spinning) electrons in a measured sample.

Fig. 13.2. The emission of left-handed and right-handed electrons in β decay. Left-handed electrons are found to predominate experimentally.

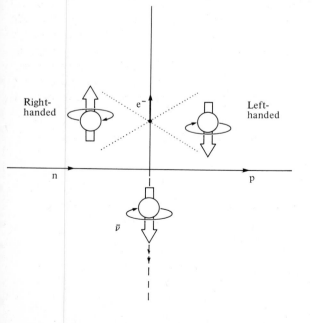

When $P = 1$, all the electrons are right-spinning and when $P = -1$, all the electrons are left-spinning. Assuming that the final-state proton does not recoil, the polarisation can be calculated to be equal to minus the ratio of the electron's speed to that of light.

$$P = -\frac{v_e}{c}.$$

So when the electrons are emitted slowly, $v_e \approx 0$ and there is no net polarisation. But when the electrons are emitted relativistically, $v_e \approx c$, they are nearly all left-handed. In 1957, F. Frauenfelder and his colleagues observed the polarisation of the electrons from the β decay of Co^{60} by scattering them through a foil of heavy atoms. They found a net polarisation of $-.4$ for electrons travelling at $.49c$, which is taken as a satisfactory agreement with the prediction.

13.4 Neutrino helicity

The four-fermion interaction constrains, very tightly, the spins which can couple through the weak interactions. In fact, when we look at the neutrino spins which are allowed to couple through the Γ we have specified, we find that only left-handed neutrinos and right-handed antineutrinos can take part in the weak interactions. As the

Fig. 13.3. Only left-handed helicity neutrinos (and right-handed antineutrinos) exist. Helicity is the component of a particle's spin along its direction of motion. It is simply a convenient way of defining the spin of moving particles.

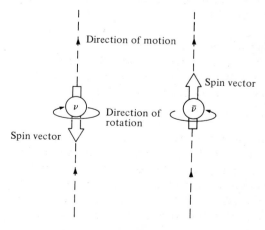

neutrinos interact only by the weak interactions, only these two possible cases are ever seen, Figure 13.3. This is a consequence of the masslessness of the neutrinos which prevents them from ever changing their spins.

To measure the neutrino helicity we must look for a particularly simple example of β decay and deduce it from the helicities of the other decay products, using the principle of angular momentum conservation.

Such an experiment was performed by M. Goldhaber and his colleagues in 1958. The spin-0 nucleus Eu^{152} is observed to undergo hybrid β decay in which an electron is captured and a neutrino emitted, leaving an excited state of the nucleus Sm^{152}. This then decays to its spin-0 ground state by the emission of a photon, Figure 13.4.

$$Eu^{152} + e^- \rightarrow Sm^{*152} + \nu$$
$$\hookrightarrow Sm^{152} + \gamma.$$

Fig. 13.4. The helicity of the photon emitted in the hybrid decay of Eu^{152} shows the single-handedness of the neutrino.

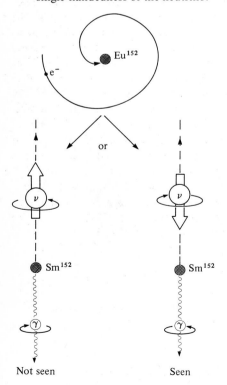

The initial and final states of the nuclei are spin 0, so if there is no angular momentum between the neutrino and the photon, then their spins must be opposite. By observing the photon helicity, that of the neutrino can be inferred, and it is invariably found to be negative (i.e. left-handed).

13.5 In conclusion

We can say that these experiments conducted in the late 1950s lend support to Fermi's original ideal of a four-fermion interaction acting at a single point, and that the phenomena of parity violation can be incorporated by choosing the interaction factors Γ to accommodate the observed spin effects, e.g. left-handed neutrinos (right-handed antineutrinos) coupling predominantly to left-handed electrons.

Other reactions went on to confirm this picture. In particular, the purely leptonic decay of the muon is an obvious candidate for Fermi's description as a four-fermion interaction, see Figure 13.5.

$$\mu^\pm \rightarrow e^\pm + \nu + \bar{\nu}.$$

The amplitude which describes this reaction is the same as that for the β decay of the free neutron, with the appropriate wavefunctions substituted. Thanks to this, it is possible to establish that the value of G_F, which is necessary to account for the observed rate of muon decay, is equal to within 2% of the value needed to account for neutron β decay. In this way we are sure that it is the same force that is responsible for these two very different processes.

The weak decay of the pion, see Figure

Fig. 13.5. The four-fermion picture of the decay of the muon.

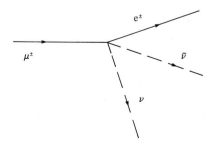

13.6(*a*), may at first be thought not to fit into the four-fermion description, viz.,

$$\pi^\pm \rightarrow \mu^\pm + v,$$

or

$$\pi^\pm \rightarrow e^\pm + v.$$

But these can be accommodated by imagining the pion to dissociate into a virtual nucleon–antinucleon pair by borrowing energy for a time allowed by Heisenberg's uncertainty principle, so the process becomes, Figure 13.6(*b*),

$$\pi^\pm \rightarrow N + \bar{N} \rightarrow \mu^\pm + v.$$

However, this rather contrived way of describing the decay is avoided in the more recent quark picture of the weak decays of hadrons, as we will soon see.

Fig. 13.6. The weak decay of the (*a*) charged pion and (*b*) its four-fermion interpretation.

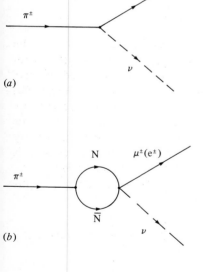

(*a*)

(*b*)

14

Two neutrinos

14.1 Introduction

Before nuclear β decay was fully understood, it was not known if the neutrinos emitted in neutron β decay were the same as those emitted in proton β decay or if they were different. (Remember that proton β decay can occur only within the nucleus, as a free proton is stable to all intents and purposes.) As the positron emitted in proton decay is the antiparticle of the electron emitted in neutron decay, it was suggested that a neutrino is emitted from one and an antineutrino from the other. This allows us to formulate a law of lepton-number conservation which was first put forward in 1953 by Konopinski & Mahmoud. If we assign a lepton number $+1$ to the electron, the negatively charged muon and the neutrino, and a lepton number -1 to the positron, the positively charged muon and the antineutrino, and a lepton number 0 to all other particles, then in any reaction the sum of lepton numbers is preserved. These assignments are summarised in Table 14.1. We can check this law in the weak interactions we have met so far:

Table 14.1. *The assignment of simple lepton number*

Particle	$e^- \mu^- v$	$e^+ \mu^+ \bar{v}$	Others
Lepton number	1	-1	0

In neutron β decay

$$n \to p + e^- + \bar{\nu},$$
$$(0) \to (0) + (1) + (-1) \checkmark.$$

In proton β decay

$$p \to n + e^+ + \nu,$$
$$(0) \to (0) + (-1) + (1) \checkmark.$$

In pion decay

$$\pi^{\pm} \to \mu^{\pm} + \bar{\nu} \ (\nu),$$
$$(0) \to (\mp 1) + (\pm 1).$$

The assignment of these lepton numbers is shown to have a physical significance by the absence of reactions which do not conserve them, but which otherwise seem feasible. For instance

$$\bar{\nu} + p \to e^+ + n$$

is observed, whereas the similar reaction

$$\bar{\nu} + n \to e^- + p$$

is not observed.

14.2 A problem in the weak interactions

The Fermi theory of the weak interactions and the lepton-number conservation law with the experiments of β, pion and muon decays formed the content of weak interaction physics until 1960. And although the theory could adequately explain the experimental observations, it could not equally well explain what was not observed. In particular, the decay of a muon into an electron and a photon was not observed despite seeming to be a perfectly valid electromagnetic transition,

$$\mu^- \to e^- + \gamma.$$

The solution to this impasse is the proposal that any neutrino belongs to either of two distinct species: one associated with the electron and one associated with the muon, and that electron-type neutrinos ν_e can never transform into muons, nor muon-type neutrinos ν_μ into electrons. So the β decay of the neutron decay involves only electron-type antineutrinos:

$$n \to p + e^- + \bar{\nu}_e,$$

and the muon decay of the pion involves only muon-type antineutrinos.

$$\pi^- \to \mu^- + \bar{\nu}_\mu.$$

The muon can decay into an electron only if a muon-type neutrino carries off the 'muon-ness' and an electron-type antineutrino cancels out the 'electron-ness' of the electron:

$$\mu^- \to e^- + \bar{\nu}_e + \nu_\mu.$$

The decay of the muon into an electron and a photon is forbidden because the 'muon-ness' is apparently transformed into 'electron-ness'.

All this means that the law of lepton conservation is now extended to that of lepton-type conservation and works in a similar fashion. Both electron-type number and muon-type number must be conserved separately in each reaction. Using the revised lepton-number assignments in Table 14.2, we can see how this works for muon decay:

$$\mu^- \to e^- + \bar{\nu}_e \quad + \nu_\mu,$$

Muon number $(1) \to (0) + (0) \quad + (1) \checkmark,$

Electron number $(0) \to (1) + (-1) + (0) \checkmark.$

Strictly speaking, each time we have written down the symbol for the neutrino in the preceding chapters, we should have associated with it a suffix denoting electron- or muon-type, a procedure we shall follow from now on.

Once again, the adoption of a new conservation law has side-stepped our difficulties (the last occasion being the introduction of strangeness in Chapter 8). It will not be the last time we adopt such an approach. Our inescapable conclusion of this conservation law is that the electron-type and muon-type neutrinos should be physically different particles, and the first modern neutrino experiment was designed to illustrate just this.

14.3 The two-neutrino experiment

Neutrino induced experiments have peculiar

Table 14.2. *The assignment of lepton-type number*

Particle	$e^- \nu_e$	$e^+ \bar{\nu}_e$	$\mu^- \nu_\mu$	$\mu^+ \bar{\nu}_\mu$	Others
Electron number	1	−1	0	0	0
Muon number	0	0	1	−1	0

difficulties due to the extremely feeble nature of the weak interactions. It is possible to use the low-energy neutrino flux from a nuclear reactor for some experiments, but others require higher-energy neutrinos which have the benefit of interacting more frequently with the targets presented. The first source of high-energy neutrinos became available in the early 1960s with the construction of the first of the big accelerators, the Alternating Gradient Synchrotron in Brookhaven in the USA. With this machine, protons could be collided into a solid target such as beryllium to produce a large flux of pions. As we have seen, these pions then decay predominantly into muons and muon-type neutrinos. It is possible to separate out just the neutrinos by passing the beam through vast quantities of iron (at thicknesses of about 20 m) to filter out the muons and any other extraneous particles, see Figure 14.1.

If there is no distinction between electron- and muon-neutrino types, then we would expect the two possible reactions to occur with equal likelihood.

$$\nu_\mu + \text{n} \rightarrow \mu^- + \text{p}, \tag{14.1}$$

$$\nu_e + \text{n} \rightarrow \text{e}^- + \text{p}. \tag{14.2}$$

However, if the two types really are distinct we would expect the first reaction to occur to the almost total exclusion of the second, as the neutrino beam consists almost entirely of muon-type neutrinos.

In the first really massive accelerator experiment of modern physics, Leon Lederman and his colleagues showed that the first (muon) reaction does indeed predominate. In 25 days of accelerator time, some 10^{14} neutrinos traversed their spark chamber which produced just 51 reactions resulting in a final-state muon. The ratio of the electrons to muons produced was later measured at CERN to be .017 ± .005, so conclusively demonstrating the existence of two separate neutrino types. This experiment demonstrated the validity of lepton-type conservation and explained the observed absence of decays which would otherwise be allowed.

Fig. 14.1. Schematic diagram of the two-neutrino experiment.

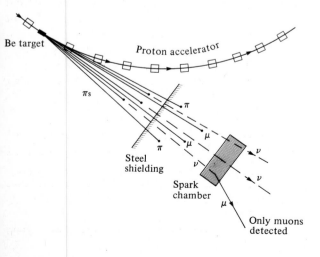

15

Neutral kaons and CP violation

15.1 Introduction

Soon after the observation of the weak interaction's violation of parity, it was discovered that it does not preserve charge conjugation symmetry **C** either. This was demonstrated by examining the spins of the electrons and positrons emitted in the decays of positively charged and negatively charged muons respectively. But physicists hoped that these two symmetry violations cancelled each other out exactly, so that the combined **CP** symmetry would be preserved by the weak interactions. To test this it is necessary to define an elementary particle state which is either even or odd under the **CP**-symmetry operation, allow the weak interaction to act, and then check that the final result has the same **CP** symmetry. It is possible to assign even or odd parity to an elementary particle state because the nature of the state remains unchanged under the parity operation, the only effect being the possible change in sign of the state wavefunction for a state of odd parity. Unfortunately, it is not possible to assign a well-defined **CP** symmetry to the K^0. This is because the operation always transforms it into its antiparticle and so changes the identity of the wavefunction. To compare the nature of the wavefunctions before and after an intended symmetry operation we must at least be sure they represent the same particle. We phrase this technically by saying that the K^0 is not an eigenstate of the **CP** operation.

Now, as we believe **CP** to be a good symmetry (one that is preserved by the weak interaction), it

follows that to describe the interaction satisfactorily we should consider it as acting on states which have a well-defined value of **CP**, i.e. on the eigenstates of **CP**. As K^0 and $\overline{K^0}$ are not these eigenstates it means that the weak interaction does not really see these particles, but some others instead which are eigenstates. The simplest of these are simple mixtures of the original particles:

$$K_1^0 = \frac{1}{\sqrt{2}} (K^0 + \overline{K^0}),$$

$$K_2^0 = \frac{1}{\sqrt{2}} (K^0 - \overline{K^0}).$$

The wavefunctions of these two states keep their identity under the **CP** operation; the K_1^0 has even **CP** symmetry and the K_2^0 odd **CP** symmetry. The weak interaction acts on these states, not on the neutral K mesons produced by the strong forces.

15.2 What is a neutral kaon?

The answer to this question depends on the interaction by which the particle is observed. The kaon which is produced in the strong interaction is either K^0 or $\overline{K^0}$, both of which are eigenstates of the parity operation with odd intrinsic parity and which have definite assignment of strangeness. The 'particle' which decays by the weak interaction is K_1^0 or K_2^0 which are eigenstates of the combined **CP** operation, but which do not have a well-defined strangeness quantum number.

Proof that the K_1^0 and K_2^0 are as real particles to the weak interaction as are K^0 and $\overline{K^0}$ to the strong can be found from their decays. The K_1^0 which is even under **CP** symmetry can decay only to states which are also even, such as a state of two pions. But K_2^0 which is odd under **CP**, can decay only to **CP** odd states, such as a state of three pions. This gives rise to very different mean lifetimes of the two particles.

$$K_1^0 \rightarrow 2\pi \qquad \tau = 0.8 \times 10^{-10} \text{ s},$$

$$K_2^0 \rightarrow 3\pi \qquad \tau = 5.2 \times 10^{-8} \text{ s}.$$

Another remarkable distinction between K_1^0 and K_2^0 is that they have different masses, although they are both equal mixtures of K^0 and $\overline{K^0}$ which have identical masses. This seemingly paradoxical conclusion was reached in 1961 when the mass difference was measured experimentally by a method which neatly shows up the identity crisis suffered by neutral K mesons.

When a neutral K meson is first produced by the strong interaction, it is definitely either K^0 or $\overline{K^0}$ depending on the strangeness of the reaction, e.g.

$$\pi^- + p \rightarrow \Lambda^0 + K^0. \tag{15.1}$$

Immediately, at the point of creation, the K^0 is an equal mixture of K_1^0 and K_2^0. However, we know that K_1^0 has a much shorter mean lifetime than K_2^0 and so at longer times from its creation the more likely it is to be a K_2^0. When the time elapsed is much greater than the mean lifetime of the K_1^0 we can say that the kaon is almost entirely K_2^0. This means that, according to the equation

$$K_2^0 = \frac{1}{\sqrt{2}}(K^0 - \overline{K^0}),$$

what originally started out as a particle, now has a 50% chance of being its antiparticle. We can see this transformation explicitly by using the kaon produced from the reaction above to undergo the reaction which produces hyperons

$$\overline{K^0} + N \rightarrow \Lambda + \pi. \tag{15.2}$$

Because of strangeness conservation, this can occur only with $\overline{K^0}$ but not with K^0. So if we take the neutral kaon in (15.1) and wait for the K_1^0 content to drop, we should start to see reaction (15.2) occur when a suitable target is placed in the beam. The frequency of reaction (15.2) will depend on the fraction of $\overline{K^0}$ which is generated in the beam by the K_1^0 component decaying. The intensity of the $\overline{K^0}$ component of the beam can be plotted as a function of time and it turns out that the nature of the variation depends on any mass difference which exists between K_1^0 and K_2^0. Experiments indicate the existence of a mass difference of about 7×10^{-6} eV. Bearing in mind the mass of K^0 of 498 MeV, the mass difference between K_1^0 and K_2^0 is about one part in 10^{14}!

15.3 Violation of CP symmetry

In 1964, Christenson, Cronin, Fitch and Turlay decided to check that the weak interaction did conserve **CP** symmetry exactly, and so to justify the use of states K_1^0 and K_2^0 as the particles appropriate to the weak force. They chose simply to observe a beam of K_2^0 and look for any decay into just two pions. If any were observed, then this would mean that the K_2^0 particle with **CP** $= -1$ had transformed into the two-pion state with **CP** $= +1$ and that the weak interaction does not conserve the symmetry. In the experiment the beam was allowed to travel about 18 m to ensure as few K_1^0 present as possible. The products of the particle decays of the K_2^0 beam were then observed as they left their tracks through the particle detectors, which also measured their energies, see Figure 15.1. They observed just a few of the forbidden decays of the K_2^0 beam into pairs of oppositely charged pions: about 50 out of a total of 23 000 decays. This was far higher than any background event rate which may have resulted from the accidental presence of K_1^0 particles still in the beam, and the team concluded that the K_2^0 could decay into just two pions and that **CP** symmetry was not preserved exactly by the weak interaction after all.

Because of this **CP**-violating effect, it follows that K_1^0 and K_2^0 are not the exact eigenstates of **CP** and so are not quite the particle 'seen' by the weak interaction. Instead, the particles, as 'seen' by the weak interaction, are basically the **CP** eigenstates K_1^0 and K_2^0, but each with a small admixture of the other. These are called the long-lived K_L, and short-lived K_S, kaons, respectively, where,

$$K_L = K_2^0 + \varepsilon K_1^0, \quad K_S = K_1^0 + \varepsilon K_2^0,$$
$$\varepsilon \approx 2 \times 10^{-3}.$$

The violation of **CP** has deep theoretical consequences which are not properly understood even now. In most weak interaction theories, **CP** is

Fig. 15.1. A schematic drawing of the **CP**-violation experiment of Christenson *et al.* (1964). Any decay of K_2^0 into just two pions represents the violation of **CP** symmetry.

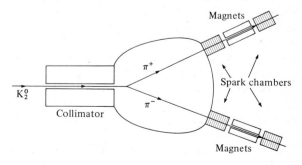

normally treated as being exact, with the violation appearing as some small perturbation to an otherwise **CP**-conserving force. One possible explanation is that the weak interaction does conserve **CP**, and that the violation is due to some 'super-weak' force which is an entirely new interaction. One intellectually satisfying consequence of **CP**-symmetry violation is that we can at last convey to our intelligent alien the absolute distinction between left and right. Violation of **CP** symmetry gives rise to an observable difference in the probabilities of occurrence (or branching ratios) of the reactions:

$$K_2^0 \rightarrow \pi^+ + e^- + \bar{\nu}_e,$$
$$K_2^0 \rightarrow \pi^- + e^+ + \nu_e.$$

We can now communicate that we define the neutrino by specifying the branching ratio of the reaction in which it is present. This establishes a common matter–antimatter convention which allows our alien to identify uniquely our handedness convention.

Part 4
Weak interaction physics II

The current–current theory of the weak interactions

16.1 Introduction

In Part 3 we examined some of the processes of the weak interactions and learnt some of their physical attributes (relatively long lifetimes of weak decays, parity-violating effects, etc.). What we want to do now is to introduce a framework which relates the very disparate phenomena of the weak force, ranging from nuclear β decay and muon decay to high-energy neutrino collisions with matter. Because some of the processes involve hadrons, it will be necessary to ensure that our framework incorporates the consequences of the $SU(3)$ internal symmetry scheme of the hadrons and desirable that it can accommodate the hypothetical quarks as the origin of this symmetry.

We start the construction of the framework by dividing the weak interactions into three classes reflecting the categories described above:

(1) *leptonic reactions* involving only leptons, such as muon decay, $\mu^- \rightarrow e^- + v_\mu + \bar{v}_e$;

(2) *semi-leptonic reactions* involving both leptons and hadrons, such as neutron β decay, $n \rightarrow p + e^- + \bar{v}_e$;

(3) *hadronic weak reactions* involving only hadrons such as the pionic decay of the kaons, $K_1^0 \rightarrow \pi^+ + \pi^-$.

16.2 The lepton current

The ultimate aim is to achieve a common description of all three classes of weak interactions. But we start by concentrating only on one, the leptonic reaction. What we have seen of the weak force provides us with our description. In these reactions we saw that whenever an electron-neutrino is absorbed, an electron is created; and equivalently, whenever an electron-neutrino is created a positron has to be created also. This is as a result of the laws of conservation of lepton number and lepton-type number. This means that in our description, the lepton wavefunctions must always come in pairs. Also, from our knowledge of β decay, we know that these wavefunctions must be coupled together via an interaction factor Γ which combines

the spins together in a correct, parity-violating way. We can now write down a 'lepton current' L^{W} which describes the flow of leptons during the weak interaction.

$$L^{W} = \bar{\psi}_e \Gamma \psi_{v_e} + \bar{\psi}_\mu \Gamma \psi_{v_\mu},$$
$$\bar{L}^{W} = \bar{\psi}_{v_e} \Gamma \psi_e + \bar{\psi}_{v_\mu} \Gamma \psi_\mu.$$

The second line is essentially the antiworld equivalent of the familiar processes.

We can now generate the first-order amplitudes $m^{(1)}$ of all the leptonic processes by specifying interactions between leptonic currents. In fact the simple product of the two lepton currents shown seems to generate all the reactions we see:

$$m^{(1)} \supset G_F \bar{L}^{W} L^{W}.$$

We can illustrate the currents and their couplings to

Fig. 16.1. The leptonic current L^{W} can be multiplied with its antiworld partner $\bar{L}^{W}$ to generate all the observed weak interaction of leptons.

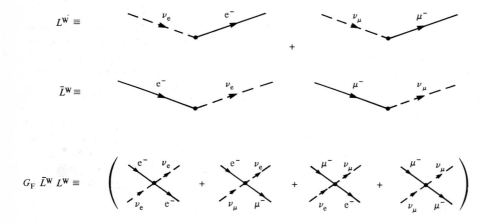

Fig. 16.2. Higher-order interactions (repeated weak interactions) can be generated by multiplying the current–current reaction with itself.

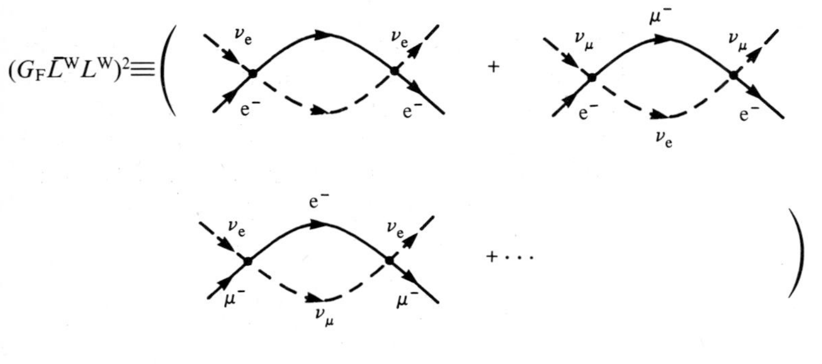

show the interactions diagrammatically, see Figure 16.1. Inspecting the interaction diagrams we must bear in mind that the destruction of a particle is equivalent to the creation of its antiparticle, so the same diagram can describe, for instance, both electron–muon scattering and muon decay.

16.3 Higher-order interactions

We can use this description to show also the higher-order interactions which can occur between the leptons. These result when the weak force acts on the leptons more than once and are described essentially by the product of two amplitudes at successive instants. For the second-order amplitudes $m^{(2)}$ this is given by the square of the simple current–current interaction.

$$m^{(2)} \supset (G_F \bar{L}^W L^W)^2.$$

These are shown diagrammatically in Figure 16.2. In fact, these higher-order interactions are not of practical importance for the weak interaction. This is because the second-order amplitudes are proportional to G_F^2 and the nth-order amplitudes are proportional to G_F^n.

Because the weak interaction is weak, the Fermi coupling constant G_F is small (compared to 1) and the higher powers of G_F are even smaller:

$$1 \gg G_F \gg G_F^2 \gg \ldots \gg G_F^n.$$

So to achieve the accurate description of a process, only the first-order term is significant. But the higher-order terms are of theoretical significance. It is desirable in principle that they should be calculable, and it was this motivation which has led us to the most recent theory of the weak force, as we shall soon see in Part 5.

17

An example leptonic process: electron–neutrino scattering

17.1 Introduction

Elastic electron-neutrino–electron scattering provides us with an example of the weak interaction at its simplest. The amplitude for the process is found in the current–current interaction.

$$m^{(1)}(\nu_e + e^- \rightarrow e^- + \nu_e) = G_F(\bar{\psi}_e \Gamma \psi_{\nu_e})(\bar{\psi}_{\nu_e} \Gamma \psi_e).$$

By inserting the mathematical expression for the wavefunctions and interaction factors into the amplitude it is possible to calculate the cross-section for the process in the laboratory frame of reference σ_{LAB}. The final answer is of a particularly simple form when the incoming neutrino energy E_{ν_e} is large:

$$\sigma_{LAB}(\nu_e + e^- \rightarrow e^- + \nu_e) = \sigma_0 \frac{E_{\nu_e}}{m_e},$$

where σ_0 is a constant factor arising from the calculation with a value of

$$\sigma_0 = 9 \times 10^{-45} \text{ cm}^2.$$

This is the very tiny effective area of interaction between the neutrino and an electron. So it is easy to understand why neutrino interactions are so rare. We can perform a very similar calculation to work out the cross-section for antineutrino–electron scattering. The answer as we might expect, is very similar at high energies:

$$\sigma_{LAB}(\bar{\nu}_e + e^- \rightarrow e^- + \bar{\nu}_e) = \frac{\sigma_0}{3} \frac{E_{\nu_e}}{m_e},$$

where we shall later see how the difference of a factor of 3 between the two results from the different-handedness of neutrinos and antineutrinos.

The fact that these cross-sections rise linearly with the energy of the incoming neutrino presents us with an almost ironic situation. When the neutrino energy is low, say $E_\nu \approx 5$ MeV, which corresponds to E_ν/m_e of about 10, then the cross-section remains around 10^{-43} cm^2. This is a minute cross-section, even for the microworld. The neutrinos which emerge from nuclear reactors are of about this energy, and so experiments using them to observe these neutrino–electron collisions must use a very high flux of neutrinos and must be prepared to wait a very long time to gather enough observations. However, when the neutrino energy is higher, say around 5 GeV (corresponding to E_ν/m_e of about 10 000) then the cross-section rises to around 10^{-40} cm^2 and the reactions are, in principle, much more accessible. Unfortunately, the only such high-energy neutrinos so far produced are from the decay of pion beams in the high-energy accelerators. Not only does this mean that the neutrino flux is limited to a rather meagre level, but also that they will nearly all be muon-type neutrinos. We can see from Figure 16.1 that elastic muon-neutrino–electron scattering is not included in the basic current–current interaction, and so such a process, if in fact it does exist, could still not provide data to test our answers. (In fact, examples of this class of process, the so-called 'neutral current' reactions, have been discovered and they necessitate modifications to the current–current interaction; we shall discuss this further in Part 5.)

So we are forced back to waiting for rare events involving neutrinos from nuclear reactors to check our answers for the cross-sections. At present, the results of the experiments can be summarised by saying that they are not inconsistent with the cross-section having the predicted levels. At worst, however, the same results do not conclusively prove that the processes exist at all.

17.2 The role of the weak force in astrophysics

Electron-neutrino–electron scattering processes may play a role in astrophysics by allowing substantial numbers of $(\nu_e, \bar\nu_e)$ pairs to transfer energy from the interiors of stars to their outer layers. Normally, photons might be thought of as fulfilling this role, but in a stellar environment they are absorbed too quickly to perform an effective transfer. So it is left to the more weakly interacting neutrinos. When a heavy star has finished burning all its hydrogen (this is fusing hydrogen nuclei together with a release of energy), it moves on to a stage in which helium is burnt (a blue giant) and after this it burns carbon (a red giant), each stage being hotter than the last. Now the existence of the neutrino energy transport effect causes the heat to be transferred throughout the star more efficiently and this results in a shorter red giant phase than would otherwise be the case. As a consequence of this we expect a larger ratio of blue giants to red giants than in the absence of the neutrino generating processes.

Observation of this ratio of stellar populations, when combined with various other astronomical and astrophysical observations, allow us to put crude limits on the strength of the weak interaction coupling constant governing the leptonic interactions involving only electron types, denoted $G_{\nu_e e^-}$. We find

$$\frac{G_F}{10} < G_{\nu_e e^-} < 10\, G_F,$$

where G_F is the coupling measured in muon decay which governs the leptonic interaction involving both electron and muon types. So we must conclude that, owing to the experimental difficulties and observational uncertainties in measuring ν_e, e^- interactions, we are unable to be sure of the identity $G_F \equiv G_{\nu_e e}$. The best we can say is that the hypothesis is not inconsistent with data.

18

The weak interactions of hadrons

18.1 Introduction

Having written down a lepton current which provided us with a description of the purely leptonic reactions, we must now extend the concept to include the semi-leptonic processes, such as nuclear β decay, which are historically more important, and also the hadronic processes. Both these categories are divided up into strangeness-conserving reactions and strangeness-changing reactions as a first step in categorising the effects of the weak interactions. Examples of the reactions in each category are shown in Table 18.1. The most obvious and serious difficulty in writing down a hadronic current (just as we wrote down the leptonic current) is that it is impracticable to write down wavefunctions for the observed hadrons; there are simply too many of them! If we were to use a separate wavefunction for each hadron, a current describing all the possible interactions would fill a book by itself. This is too complicated to be true.

18.2 The hadronic current

To proceed we can avoid the problem of wavefunctions for the hadrons and simply characterise the hadronic current in terms of its effect on the quantum numbers of the participating particles. So we can write the total weak interaction current as a sum of leptonic and hadron components.

$$J^W = L^W + H^W.$$

As before, the weak interaction amplitudes are generated by the product of the total current with its antimatter conjugate multiplied by the Fermi coupling:

$$G_F \bar{J}^W J^W = G_F (\bar{L}^W L^W + \bar{L}^W H^W + \bar{H}^W L^W + \bar{H}^W H^W).$$

which contains all the leptonic reactions ($\bar{L}^W L^W$), the semi-leptonic reactions ($\bar{L}^W H^W + \bar{H}^W L^W$), and the purely hadronic reactions ($\bar{H}^W H^W$). The form of the hadronic current (less the wavefunctions) is speculated to consist of one part which conserves the strangeness of the participating hadrons $h^\pm$ and of one part which changes it $s^\pm$

$$H^W = h^\pm \cos\theta_C + s^\pm \sin\theta_C.$$

The relative strength of the two components is governed by the Cabbibo angle θ_C which is a parameter intrinsic to the weak interactions and which must be measured from experiments.

Table 18.1. *Categories of the weak interactions of hadrons*

Reaction \ Class	Strangeness-conserving	Strangeness-changing
Semi-leptonic	$\pi^\pm \to l + \nu(\bar{\nu})$	$K^\pm \to l + \nu$
	$n \to p + e^- + \bar{\nu}_e$	$K^\pm \to \pi^0 + l + \nu$
	$\mu^- + p \to \nu_\mu + n$	$\Lambda^0 \to N + l + \nu$
	$K^0 \to K^\pm + e^\mp + \nu(\bar{\nu})$	$\Xi \to \Lambda + l + \nu$
	$\bar{\nu} + p \to e^+ + n$	$\nu + p \to e^- + \Lambda$
Hadronic	Parity-violating effects in ordinary hadron physics, e.g. $p + p \to p + p$	$K^0 \to n\pi$ $(n = 2, 3)$ $\Lambda^0 \to \pi^- p$ $\Sigma \to n\pi$

denotes $e^\pm$ or $\mu^\pm$.

At this point we must remember that just as the leptonic current L^W contains the interaction factors Γ (which are a mixture of vector and axial vector quantities), so too do the separate components of the hadronic current to ensure the correct, parity-violating coupling of hadronic spins in the reactions. So the hadronic weak current has four separate components:

(1) a vector current which conserves strangeness;
(2) a vector current which changes strangeness;
(3) an axial vector current which conserves strangeness;
(4) an axial vector current which changes strangeness.

18.3 Current algebra

Current algebra is concerned with deriving relationships between the various components of the hadronic current on the basis of the internal symmetry scheme governing the hadrons. Because of this symmetry group $SU(3)$ there exist relationships between the generators of the symmetry transformations (the group algebra) which dictate the allowed quantum-number structure of the hadrons. The fact that the components of the hadronic weak current can change the quantum numbers of the hadrons suggests they have an intimate connection with the symmetry scheme and, in fact, both the vector and axial vector components of the hadronic current form separate octet representations of $SU(3)$. This allows us to categorise the effects of the strangeness-conserving and the strangeness-changing components (each the sum of vector and axial vector parts) have on the quantum numbers of the hadrons in an interaction:

$$h^\pm \ (\Delta Y, \Delta I, \Delta I_3) = (0, 1, \pm 1),$$
$$s^\pm \ (\Delta Y, \Delta I, \Delta I_3) = (\pm 1, \tfrac{1}{2}, \pm\tfrac{1}{2}).$$

Armed with this knowledge of the effects of the weak interaction on the quantum number on the hadrons, it is possible to write down expressions for the amplitudes of the various reactions (using some imagination when it comes to the hadron wavefunctions), and to proceed to a description of the reactions.

The rate for pion decay into leptons,

$$\pi^\pm \to l^\pm + v(\bar{v})$$

can be calculated with just one unknown remaining

in the formula as a relic of the pion wavefunction. Pion β decay, which is a close relative to free neutron β decay, explores different components of the hadronic current

$$\pi^\pm \to \pi^0 + e^\pm + v_e(\bar{v}_e)$$

and, most importantly, allows the experimental determination of the Cabbibo angle. This turns out to be

$$\cos \theta_C = 0.97.$$

Inserting this into the hadronic current gives

$$H^W = 0.97h\pm +0.24 \ s^\pm,$$

which shows the strangeness-conserving component of the current to be much more important than the strangeness-changing part. Building on this information it is possible to go on to establish relationships for reactions such as nuclear β decay, neutrino–nucleon scattering and the purely hadronic decays. These predictions enjoy varying degrees of success, but all suffer the deficiency of the inadequate treatment of the hadron wavefunctions.

18.4 The hadron current and quarks

The hadron current takes on a very simple form when written in terms of the quarks. Naively we can think of the weak current as simply changing one flavour of quark (u, d or s) into another, and so changing the quantum numbers of the parent particle.

In fact, it is not quite as simple as this because the weak interaction does not specifiy a unique transformation say, from a u quark into a d quark. As we have seen, the weak current has both strangeness-conserving and -changing components which implies that the u quark can have a certain probability of transforming into a d quark, and

Fig. 18.1. The weak interaction transforms quarks.

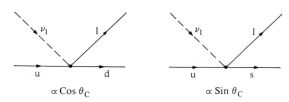

another of transforming into an s quark. So the current is written

$$H^W = \bar{u}\Gamma(d \cos \theta_C + s \sin \theta_C).$$

This current is shown symbolically in Figure 18.1. The decay of a strange meson is illustrated in Figure 18.2.

This representation is very useful for envisaging the flow of quantum numbers during a reaction. Unfortunately, it is of limited use in calculating dynamical quantities because we have very little idea how to represent the confinement of quarks mathematically. In the diagram the ignorance is contained within the shaded blobs.

Fig. 18.2. The quark picture of quantum-number flow during the semi-leptonic decay of the kaon.

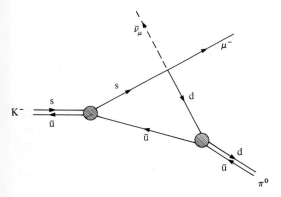

19

The W boson

19.1 Introduction

It is true to say that we can explain all the data from low-energy weak interaction processes with the Fermi theory (expressed in the framework of the current–current theory of lepton processes). Unfortunately, this theory of a point-like interaction makes unacceptable predictions for high-energy weak interactions. We have just seen how the theory predicts that cross-sections for neutrino–electron scattering will rise linearly with the energy of the incoming neutrino. But we must realise that this prediction cannot be true for arbitrarily high energies. For instance, if it were true, neutrinos with exceedingly high energies (say those in cosmic rays, originating in space) would have a very high cross-section for interacting with matter, so we would expect neutrino collisions to be commonplace events in cosmic ray photographs and more common in laboratory bubble chambers. This is just not true, and we must accept that our formula for the neutrino cross-section is valid only at small energies. Also, there are extremely well-founded theorems resting only on assumptions, such as causality, which constrain the rate at which cross-sections can rise with energy.

To solve this problem, and also to put the description of the weak interaction on a common footing with the theories of electromagnetism and the strong nuclear force, it is necessary to abandon the four-fermion point-like interaction and replace it with a particle exchange mechanism (just like, say, pion exchange between nucleons).

The particle which carries the weak interaction is called the intermediate vector boson, denoted W, see Figure 19.1. What we must do is to go back and describe all the weak interaction phenomena in terms of a particle exchange mechanism, which must approximate to the four-fermion point-like interaction at low energies to preserve its successful explanation of the data.

19.2 The W boson

The essence of the W-boson mechanism is that the lepton current no longer couples directly to itself at a single point in space–time but, instead, to the W-boson wavefunction which at some different space–time point couples with the conjugate current. The basic weak interaction amplitude $m^{(1)}$ is thus the coupling of the current with the W-boson wavefunction W see Figure 19.2.

$$m^{(1)} \equiv g(L^W \bar{W} + W \bar{L}^W),$$

$$m^{(1)} \equiv g(\bar{\psi}_e \Gamma \psi_{v_e} \bar{W} + W \bar{\psi}_{v_e} \Gamma \psi_e).$$

The first thing to note about the W boson is that it must come in both positively and negatively charged versions if it is to allow the transformations of positrons and electrons into antineutrinos and neutrinos respectively. Also, if we wish to describe neutral current phenomena, we must allow the existence of a neutral W boson as well. The role of the differing charge states is shown in Figure 19.2.

Another property of the W boson, which is easily established, is that it must be very massive. Recalling our simple argument of Chapter 7, using Heisenberg's uncertainty principle: the range of the force may be thought of as typified by the maximum

distance which its carrier can travel in the time element allowed by the uncertainty principle. The more massive the carrier, the shorter the range of the force. As the Fermi theory managed quite satisfactorily with a point-like assumption, it follows that the W boson must be more massive than the pion (which allows the strong force the measurable range of 10^{-15} m).

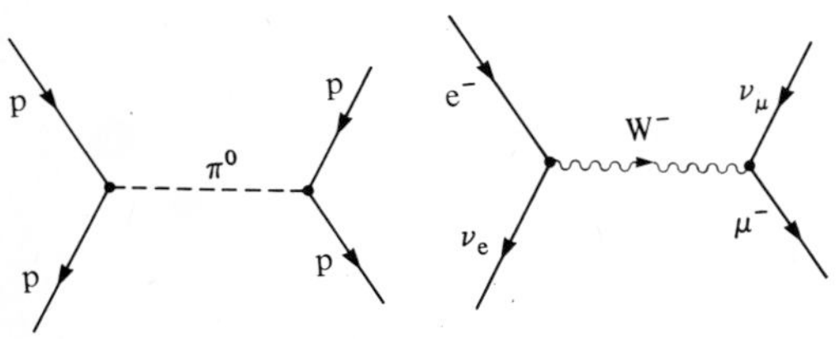

It is clear that the simple amplitudes of Figure 19.2 represent the creation (or destruction) of a W boson from the (into the) familiar leptons. The reactions involving only leptons as external particles will result from higher-order interactions represented by products of these basic amplitudes

$$m^{(2)} = g^2(L^W \bar{W} W \bar{L}^W),$$

$$m^{(2)} = g^2(\bar{\psi}_e \Gamma \psi_{v_e} \langle \bar{W} W \rangle \bar{\psi}_{v_e} \Gamma \psi_e).$$

Figure 19.3 shows the W-exchange amplitudes for some of the basic weak interaction processes which we have met so far. The factor in the amplitude describing the propagation of the W boson $\langle \bar{W} W \rangle$ is known as the W-boson propagator, which acts as its wavefunction between its creation and its destruction. It is this new factor which improves the unacceptable high-energy behaviour of total cross-sections. The mathematical expression for the W-boson propagator describes its mass and its spin (spin 1 for a vector particle) and allows us to relate

Fig. 19.2. Basic lepton processes defining the charge states required of the W boson.

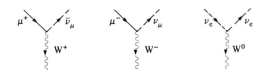

Fig. 19.1. Just as the pion carries the strong force between hadrons, so the W boson carries the weak force between leptons.

Fig. 19.3. Basic weak interaction processes mediated by W exchange.

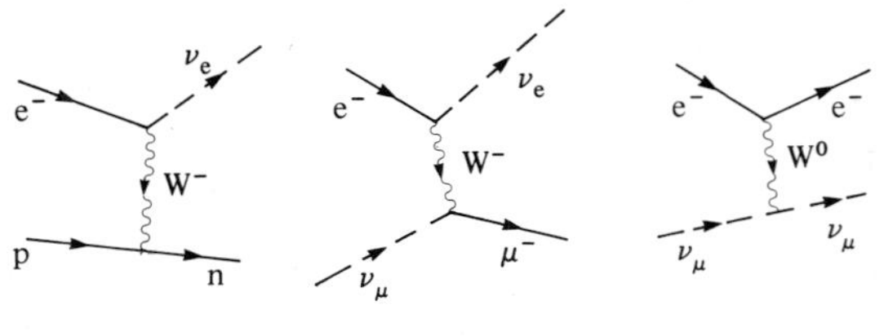

the old Fermi coupling G_F to the new lepton–W coupling g. At low energies we find:

$$G_F \propto \frac{g^2}{M_W^2} .$$

Knowing the expression for the W propagator as well as the wavefunction for the external leptons and the interaction factors allows us to recalculate the cross-sections for all processes of interest. At low energies, the answers are the same as for the Fermi theory, as desired.

19.3 Observing the W boson

We have suggested that the W boson will be spin 1, just like the photon, come in three charge states and will decay into the familiar leptons. Its discovery was awaited for many years. During this time, it was generally accepted that its mass (unknown, but assumed to be very large) prevented its manifestation in any of the experiments by then conducted. But, as we will see in the next part, the W-boson mass can be predicted from the modern theory of the weak interactions (when supported by relevant experimental data). Furthermore, as described in Part 9, experiments at CERN currently in progress have achieved the first detection of the W boson, just as predicted.

Part 5
Gauge theory of the weak interactions

Motivation for the theory

20.1 Introduction

The description of the weak interactions of leptons afforded by the current–current theory of Chapter 16 provides a good account of the experimental observations. This description includes the use of wavefunctions for the leptons (possible because there are just a few of them) and can also incorporate the use of W bosons in the role of 'the photons of the weak interactions'. We might then wonder why it is not possible to write down a fully fledged theory of the weak interactions of leptons mediated by the W bosons, similar to the QED theory of electrons and photons. Although there are no pressing practical reasons for doing so, there is considerable theoretical motivation. Firstly, it would be satisfying to be able to calculate answers to any desired degree of accuracy, in contrast to using only the simplest interactions to which our current understanding of the W bosons limits us. Secondly, we would like to have a predictive theory which could demonstrate its relevance by actually revealing new phenomena. And finally, if we can derive a theory similar to QED, this may allow us to formulate a unified theory of weak and electromagnetic forces.

20.2 Problems with the W bosons

Given the motivation and the basic building blocks of the theory, we immediately discover several problems arising in the use of the W bosons. These problems originate in the fact that these particles have spin 1 *and* non-zero mass. Let us investigate the consequences of this seemingly innocuous combination.

When we first talked about quantum-mechanical spin, we noted that although the naive picture of the spinning ball is helpful, it is a simplification of a more fundamental attribute. In fact, the spin of particles is the method for categorising their transformation properties under Lorentz transformations. A spin-1 particle is said to transform like a vector, which means that it has three components to define its orientation (i.e. its polarisation) at any point. Normally, we define the components such that two (transverse) are perpendicular to and one (longitudinal) is parallel to the direction of the momentum of the particle, see Figure 20.1. This is appropriate for a *massive* spin-1 particle, but for a *massless* spin-1 particle, like the photon, the transformations of special relativity show that the longitudinal degree of freedom has no physical significance. The photon must always be polarised in the plane transverse to its direction of motion. The difference becomes important when the propagators carry very high momentum. The two-component propagator for the photon becomes a dimensionless number depending only on the value of the momentum. But the massive vector particle must describe its mass by the presence of a mass factor contained in the extra longitudal component of its propagator, and this causes trouble.

We have seen that, in perturbation theory, the probability of an event occurring is given by the sum of contributions representing increasingly complex Feynman diagrams. Each of these contributions should be, to all intents and purposes, a simple dimensionless number. However, the W-boson propagator has its dimensional mass factor which spoils its contribution. To correct this, each propagator (for each wriggly line) must be multiplied by corresponding factors of momentum to give a dimensionless contribution. However, as we have seen in QED, it is necessary to sum over all the unobserved, internal momenta to account for all the possible configurations of internal momentum a given graph may take. The presence of momentum factors multiplying each of an arbitrary number of propagators means that the mathematical expressions for the increasingly complicated diagrams will become infinite as each is summed over all possible internal momenta. The diagrams are said to diverge.

In summary, it is the mass factor in the W-boson propagator which leads to an ever-increasing number of infinite contributions to the perturbation series. It is not possible for these to be reabsorbed into redefinition of the masses and couplings. The theory is unrenormalisable and

Fig. 20.1. Vector propagators. A massive vector particle can have three components of polarisation (*a*), a massless vector particle only two (*b*).

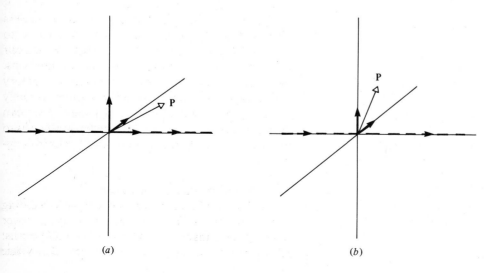

(a)   (b)

incapable of providing sensible answers to an arbitrary degree of accuracy.

Related to the problem of the theory's lack of renormalisability is the bad high-energy behaviour exhibited by some processes involving the W bosons. The presence of the mass factor in the W-boson propagator causes the cross-sections for these processes to rise with energy faster than is allowed by fundamental theorems, see Figure 20.2. This bad high-energy behaviour is precisely the problem which the W bosons were introduced to cure! As it is the mass of the W bosons which seems to be the source of the trouble, the best thing for us to do is to study the origins of these particles and their mass further.

Fig. 20.2. A process with bad high-energy behaviour caused by the mass of the W bosons.

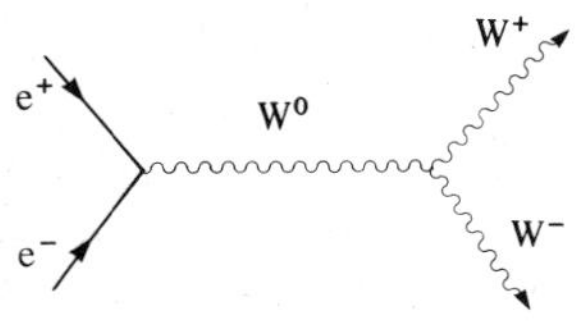

21

Gauge theory

21.1 Introduction

The principle of gauge invariance is perhaps the most significant of the concepts used in modern particle theories, as it attempts no less than to describe the origin of the forces themselves. It appears to apply to all of the four known forces described in Chapter 5 (in one guise or another) and so may eventually provide us with the basis for a comprehensive unified theory. The basic method of gauge theory is to ensure that the Lagrangian describing the interaction of particle wavefunctions remains invariant under certain symmetry transformations which reflect conservation laws observed in nature. As a first step, we can see how this works in QED.

21.2 The formulation of QED

QED seeks to explain the interaction of charged particles, say electrons, in such a way that total electric charge is always conserved. To represent this, the Lagrangian which describes the electron wavefunction must be invariant under a certain group of symmetry transformations **G**. We write this symbolically:

$$\mathbf{G}\mathscr{L}(\psi_e) \to \mathscr{L}(\psi'_e).$$

In fact, the group in question, denoted $U(1)$, is simple shift in the phase of the electron waves. This is called a global phase, or gauge, transformation because it represents an identical operation at all points in space. It is a relatively simple symmetry and it does not place a very close restriction on the form of the Lagrangian. The exercise becomes more interesting when we examine the symmetry of the Lagrangian under gauge transformations which vary according to position,

$$\mathbf{G}(x)\mathscr{L}(\psi_{\mathrm{e}}) \to \mathscr{L}'(\psi'_{\mathrm{e}}).$$

Such a local gauge transformation represents a space-dependent convention defining the phase of the electron wave. Because of the space dependence, the Lagrangian is changed by the transformation and the theory is not symmetric under this more demanding symmetry. However, it comes as a pleasant surprise to find that the presence of the photon restores such symmetry.

So far, we have mentioned only electrons interacting at a point. Of course, we know that, properly, this should be described by the electrons interacting with a photon at one point, the propagation of the photon and its subsequent interaction with other electrons at another space–time point. It so happens that the changes in the photon wavefunction cancel out the changes in the Lagrangian during a local transformation and so its introduction restores the invariance. We write this symbolically,

$$\mathbf{G}\mathscr{L}(\psi_{\mathrm{e}}, A) \to \mathscr{L}(\psi'_{\mathrm{e}}, A').$$

where the symbol A denotes a four-vector describing the electromagnetic potential (i.e. the photons' wavefunctions).

So invariance of the Lagrangian under local gauge transformations requires the existence of the photon: the long-range electromagnetic field. A physical explanation of its existence may be thought of as communicating the different space-dependent conventions defining the phase of the electron wave between different points in space.

An interesting fact is that the presence in the Lagrangian of a term attempting to describe a hypothetical mass for the photon would destroy the gauge invariance. As we have already seen, massive spin-1 particles generally give rise to non-renormalisable theories. So we may suspect local gauge invariance to be a good guide to the renormalisability of theories.

21.3 Generalised gauge invariance

Our next task is to generalise the principle of gauge invariance to other particles and other forces. The most convenient example is one which in fact turned out to be untrue, but which follows on naturally from our previous discussions.

The charge independence of the strong nuclear force means that it acts identically on both proton and neutrons. It cannot distinguish between them but instead 'sees' only one basic nucleon N. This led us to categorise the proton and neutron as the two isospin components of the isospin-$\frac{1}{2}$ nucleon. The charge independence of the force may then be expressed as its invariance under rotations in isospin space, and associated with the invariance is the conservation of the total isospin of the system on which the forces act. The Lagrangian which describes the interaction of nucleons should then be invariant under the gauge group of isospin transformations $SU(2)$,

$$\mathbf{G}^{SU(2)}\mathscr{L}(\mathrm{N}) \to \mathscr{L}(\mathrm{N}'),$$

where the group $SU(2)$ effectively rotates a proton into a neutron and vice versa. As the strong force cannot distinguish between the two, any possible mixture of the two can be used to define a nucleon. The global transformation essentially redefines the nucleon convention at all points in space.

As before, it is possible to require that the theory be symmetric under the more demanding local gauge transformations and, as before, it is found necessary to introduce a gauge particle ρ to ensure the invariance of the Lagrangian:

$$\mathbf{G}^{SU(2)}(x)\mathscr{L}(\mathrm{N}, \rho) \to \mathscr{L}(\mathrm{N}', \rho').$$

The source of this new gauge particle is isospin, just as the source of the electromagnetic gauge field is electric charge. Because the nucleon may or may not change its electric charge in interaction, the new gauge particle must come in three charge states to do its job, (ρ^+, ρ^0, ρ^-). Furthermore, because electric charge is related to isospin by $Q = e[I_3 + (Y/2)]$, the ρ gauge particle carries its own isospin and so can act as its own source. This allows the charged gauge particle to interact with itself, in contrast to the neutral photons in QED, see Figure 21.1. The difference between the two is expressed by saying that QED is an Abelian field theory whilst the $SU(2)$ gauge theory is non-Abelian. The difference is a consequence of the mathematical structure of the gauge groups. The simple shift in phase is said to be Abelian because a series of transformations can be performed in any order to produce the same effect as one big transformation. The group of rotations in isospin space $SU(2)$ (which is also the group of ordinary spatial rotations in the 3-dimensional space of the everyday word) is non-Abelian because a series of transformations does depend on the order of operation.

Gauge invariance was first generalised to the isospin invariance of the Lagrangian by Yang and Mills in 1954 and Shaw in 1955, but further work was deemed nugatory as such invariance apparently required the existence of a charge triplet of massless spin-1 gauge bosons (the ρ particles) which do not exist. A few years later the ρ mesons were discovered and the possibility of a gauge theory for the strong interaction was discussed. Unfortuntely, the ρ mesons are massive bound states of two pions and cannot be considered as candidates for this fundamental rôle.

21.4 Gauge invariance and the weak interactions

The first step we must take in formulating a sensible theory is to incorporate the basic laws of lepton physics, which we have already met. These are the separate laws of electron number and muon number. Because of them, it is natural to group the leptons' wavefunctions into families of the same lepton number,

$$l_e = \begin{pmatrix} \nu_e \\ e^- \end{pmatrix} \quad l_\mu = \begin{pmatrix} \nu_\mu \\ \mu^- \end{pmatrix}.$$

Fig. 21.1. In QED, photons cannot interact directly, only with electrons (*a*). In SU(2) gauge theory, the gauge particles can interact both directly and indirectly (*b*).

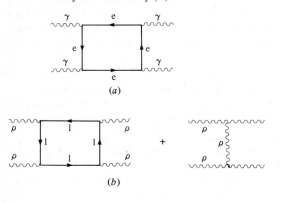

(a)

(b)

If gauge theory is now to be used, it is necessary to identify some conservation law which will imply the existence of some symmetry of the Lagrangian. To discover this, we note the similarity of the doublets to the isospin doublet of the nucleon containing the proton and the neutron. We can then use this similarity to propose that the weak interaction also is independent of the electrical charge of the particles on which it acts. The weak intraction 'sees' only a lepton and cannot distinguish between a neutrino and an electron.

This leads us to define 'weak isospin' in a fashion exactly analogous to the isospin of nucleons, and allows us to require that the weak interaction be invariant under rotations in this weak isospin space. So the Lagrangian must be invariant under the group of isospin rotations, denoted by $SU(2)^W$ to show it is acting on the leptons' wavefunctions,

$$\mathbf{G}^{SU(2)^W} \mathcal{L}(l_e, l_\mu) \to \mathcal{L}(l'_e, l'_\mu).$$

In exact parallel to the previous discussion, enforcing the more demanding local symmetry requires the introduction of gauge particles to ensure invariance of the Lagrangian:

$$\mathbf{G}^{SU(2)^W}(x) \mathcal{L}(l_e, l_\mu, W) \to \mathcal{L}(l'_e, l'_\mu, W').$$

Here the physical reason for the existence of the W boson is to communicate between interacting leptons the locally defined convention governing the mixture of 'electron' and 'neutrino' constituting the lepton. Again, the gauge particle must be a charge triplet (W$^+$, W^0, W$^-$), and here we have something new: the presence of the W^0 particle will allow neutral current reactions in which a neutrino does not have to become an electron, see Figure 21.2. But this theory was first discussed well before the discovery of neutral currents and the requirements

Fig. 21.2. The neutral gauge particle allows weak neutral current reactions.

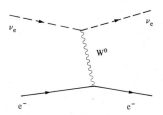

for a W^0 seemed a positive inconvenience. Also inconvenient is the fact that gauge invariance allows only massless gauge particles in comparison with the large minimum mass of the W boson required by lack of experimental observation.

These two factors hampered the development of gauge theory for almost a decade. What was required was a mechanism by which the W bosons may be allowed to have mass whilst originating from a gauge-invariant (and thus possibly renormalisable) theory. Also, the observed absence of neutral currents withheld the experimental confirmation of the theory necessary for its credibility and its development.

22

Spontaneous symmetry breaking

22.1 Introduction

Obtaining asymmetric solutions (i.e. massive W bosons) from a symmetric theory (i.e. gauge-invariant theory) is common to many branches of physics. For example, the magnetic field of a magnet obviously defines a preferred direction in space (i.e. it has broken rotational symmetry), but the equations governing the motions of the individual atoms are entirely symmetric. How has this come about? The answer lies in the fact that the symmetric state is not the state of minimum energy, and that in the process of evolving towards the minimum energy state, the system has broken its intrinsic symmetry. A simple mechanical example is the behaviour of a marble inside the bottom of a wine bottle, see Figure 22.1. The symmetric state is obviously in the middle on top of the hump, but this is not the state of least energy, as the marble possesses potential energy due to its elevation. A small perturbation will send the marble tumbling into the trough where the system will possess least energy but will also be rotationally asymmetric.

22.2 Spontaneous breaking of global symmetry

We now want to apply these ideas to particle theory to see if a broken gauge symmetry has anything to do with particle mass.

Let us start by considering a hypothetical particle wavefunction consisting of two components (just as we consider the nucleon wavefunction to consist of proton and neutron component).

$$\phi \equiv (\phi_1, \phi_2) \quad \text{just as} \quad N \equiv (p, n).$$

We can write down a Lagrangian which specifies the energy of the interaction between ϕ_1 and ϕ_2 in an equally hypothetical fashion. We may even choose the wine-bottle shape of the previous section, see Figure 22.2. The ϕ_1 and ϕ_2 axes are the values of the particle wavefunctions (which are related to the probability of finding the particle at that point). For the interaction we have chosen, the energy is not a minimum at zero values for the wavefunctions, but around the circle

$$\phi_1^2 + \phi_2^2 = R^2.$$

Despite this unusual property, the Lagrangian is still symmetric under transformations between ϕ_1 and ϕ_2, i.e. rotations in the ϕ_1, ϕ_2 plane,

$$\mathbf{G}\mathscr{L}(\phi_1, \phi_2) \rightarrow \mathscr{L}(\phi_1', \phi_2'),$$

Fig. 22.1. The initial position of the marble is symmetric but not minimum energy. A small perturbation will cause the rotational symmetry to be broken and the system to assume the state of minimum energy.

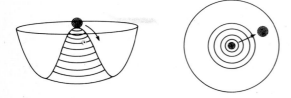

Fig. 22.2. The interaction energy chosen for the two components of a hypothetical wavefunction. The state of minimum energy corresponds to a non-zero value for the wavefunction.

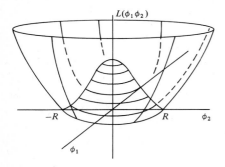

where for the moment we are dealing with global transformations: $\mathbf{G}$ is the same at all points in space.

For convenience, we might like the state of minimum energy to be that in which the values of the wavefunctions are zero. In fact, this is necessary if we want to use the mathematics of perturbation theory. We can arrange this by a simple redefinition of the wavefunctions.

$$\phi_1' = \phi_1, \quad \phi_2' = \phi_2 + R.$$

This simply draws new axes through the point R in the plane. If we write the Lagrangian in terms of the new wavefunctions, it should describe exactly the same physics; after all, all that we have done is to redefine the particles:

$$\mathscr{L}(\phi_1, \phi_2) \equiv \mathscr{L}(\phi_1', \phi_2').$$

However, some interesting features arise in this redefined system. Firstly, the new Lagrangian is not invariant under the original group $\mathbf{G}$ of transformations rotating ϕ_1 into ϕ_2. This is not surprising because the potential is not symmetric about R. Secondly, the expression for the new Lagrangian describes ϕ_2' as a particle of mass related to R and ϕ_1' as a massless particle. This is in contrast to the original Lagrangian in which the concept of mass was rather ill defined. The combined result is that we have broken the gauge invariance of the original Lagrangian and, as a result, given a well-defined mass to one of the particles involved, whilst the other remains massless. The interesting question now is whether or not particle masses in the real world originated in a similar fashion from some originally gauge-invariant (and hence possibly renormalisable) interaction.

Unfortunately, things are not of much use as they stand in this simple model. The presence of the massless particle turns out to be a general consequence of this type of mechanism (a theorem proved by the Cambridge physicist Jeffrey Goldstone in the early 1960s). This is unfortunate as no such massless and spinless particles seem to exist in the real world.

The particle which has developed a mass is nothing like a W boson and so it will still remain massless. Our trouble seems to be increasing! Happily, in the next section all these difficulties magically vanish.

22.3 Spontaneous breaking of local symmetry – the Higgs mechanism

If we now take the original Lagrangian representing the wine-bottle shape of interaction energy and demand that it be invariant under local gauge transformations (i.e. the rotations in the ϕ_1, ϕ_2 plane may vary from place to place), then we know from Chapter 21 that we must introduce a gauge particle to maintain the invariance:

$$\mathbf{G}(x)\mathscr{L}(\phi_1, \phi_2, A) \to \mathscr{L}(\phi_1', \phi_2', A').$$

In this instance, the gauge particle is responsible for communicating the ϕ_1, ϕ_2 content of ϕ from place to place. As before, we may redefine the wavefunctions so as to arrange our axes of zero wavefunction value to pass through the point of minimum energy and write the Lagrangian in terms of these redefined wave functions:

$$\mathscr{L}(\phi_1, \phi_2, A) \to \mathscr{L}(\phi_1', \phi_2', A').$$

It is in this final step that something remarkable occurs. The redefined ϕ_2' particle, as before, acquires a mass proportional to R, but the massless Goldstone boson ϕ_1' disappears and, instead, the formerly massless gauge particle now acquires a mass, again proportional to R.

What in fact happens is that the expressions describing the original massless gauge particle become mixed with the number (vacuum expectation value) R in such a way as to create a mass term, whilst the Goldstone boson ϕ_1' becomes absorbed into the gauge particle in such a way as to lose its physical significance.

The physical interpretation of all this is as follows. The original Lagrangian contains a two-component wavefunction ϕ_1, ϕ_2, and a massless vector gauge field which consists of two polarisation states. The redefined Lagrangian contains one massive particle ϕ_2', and a massive vector gauge field which, by virtue of its mass, now contains three polarisation states. The number of physical degrees of freedom in the system remains the same, but the mapless Goldstone boson has become the third polarisation state of the massive gauge boson. This looks more encouraging. The Goldstone bosons have been avoided by using *local* gauge symmetries, a step first taken by Peter Higgs of Edinburgh, and others, in 1964. What is more, the gauge bosons have acquired mass from an originally gauge-invariant theory; this was the point of the entire exercise. The only price to be paid for this success is the presence of the massive ϕ_2' particle, the so-called Higgs particle. This should be a real observable particle but has not been seen so far. Its observation would lend great support to the idea of spontaneous symmetry breaking.

23

The Glashow–Weinberg–Salam model

23.1 Introduction

In 1967 and 1968 respectively, Steven Weinberg of Harvard and Abdus Salam of London independently formulated a unified theory for the weak and electromagnetic interactions, based in part on work developed previously by Sheldon Glashow, also of Harvard. The theory describes the interactions of leptons by the exchange of W bosons and photons and incorporates the Higgs mechanism to generate the masses for the W bosons. Because the Lagrangian prior to spontaneous symmetry breaking (e.g. prior to the re-definition of the fields) is gauge invariant, Weinberg and Salam conjectured, although were not able to prove, that the theory is renormalisable. The proof was demonstrated subsequently by Gerard 't Hooft of Utrecht, in 1971.

The idea of the model is to write down a locally gauge-invariant Lagrangian describing the interactions of leptons with massless W-gauge bosons, just as described in Chapter 21. Hypothetical Higgs wavefunctions are then introduced with a suitably chosen interaction Lagrangian which is added to that for the leptons. Following the re-definition of the Higgs wavefunctions, the Lagrangian now describes particles with mass but is no longer gauge invariant under the same local transformations. Because the theory is renormalisable, the Feynman rules can be used to calculate finite answers for any physical quantities to any desired degree of accuracy. We must take care during the spontaneous symmetry breaking to ensure that the photon remains massless whilst the W bosons are made massive.

This is achieved by a sufficiently clever choice of the Higgs interactions such that, after redefinition of the fields, the Lagrangian is invariant under some sub-group of the local gauge transformations. Because this model has been so successful over the past decade, it deserves a more detailed examination.

23.2 Formulation

We have previously grouped the electron with the electron-neutrino as two different, weak isospin states of a single lepton wavefunction,

$$ l_e = \begin{pmatrix} v_e \\ e^- \end{pmatrix}. $$

However, a straightforward grouping like this is unsatisfactory, for while the neutrino is massless and left-handed, the massive electron is both right- and left-handed. We have already mentioned that the weak interaction prefers its electrons to be left-handed, such that if the electron were actually massless (a state well-approximated by very relativistic electrons) it would act only on left-handed electrons. Therefore, let us split the electron wavefunction into separate right- and left-handed components and group them separately

$$ l_e = \begin{pmatrix} v_e \\ e^- \end{pmatrix}_L, \quad e^-_R. $$

To do this properly, the electron can be allowed to have no mass. So we must arrange for this mass to be generated later.

We now wish to write down a Lagrangian describing the interaction of these leptons and to introduce gauge bosons by requiring the Lagrangian to be invariant under certain gauge transformations. To do this, we must know the generators of these transformations (in fact, these are the quantities which are conserved by the interactions). These quantities will differ before and after spontaneous symmetry breaking. Our hypothesis is that the present state of the world is the result of this symmetry breaking, and so we have relative freedom in choosing conservation laws prior to the breaking, provided that, after it, electric charge is conserved. The left-handed component of the electron and the neutrino are the $I_3^W = -\frac{1}{2}$ and $I_3^W = +\frac{1}{2}$ components of the lepton weak isospin doublet $I^W = \frac{1}{2}$. Because the right-handed component of the electron has no weakly interacting partner it must be isospin zero $I^W = 0$. (It must transform into itself under rotations in weak isospace.) At this stage, prior to spontaneous symmetry breaking, the two components of the lepton must be identical except for the third components of their leptonic isospin. However, the electron has negative electric charge whilst the neutrino is neutral. So we must relate this electric charge difference to the difference in their component isospin values. We can do this by introducing 'leptonic hypercharge' Y^W which is defined by the equation:

$$Q = e(I_3^W + Y^W/2).$$

So, by awarding both v_e and e_L^- unit weak hypercharge, the difference in their electric charges is given by the difference in their component isospin values. The 'weak quantum numbers' of all the leptons are shown in Table 23.1.

We now demand that the interactions between leptons conserve leptonic isospin and hypercharge. We express this by demanding the Lagrangian be invariant under the $SU(2)_L^W$ group of transformations in weak isospace and the $U(1)$ group of transformations generated by weak hypercharge (again a simple shift in the phase of the fields). To achieve these invariances under local (space-dependent) transformations, we must introduce the appropriate gauge particle:

$$\mathbf{G}^{SU(2)_L^W}(x)\mathscr{L}(l_L, W) \to \mathscr{L}(l_L', W').$$

We introduce the W-gauge particle to maintain invariance under rotations in weak isospace, and we

introduce the B-gauge particle to maintain invariance under the lepton wavefunction phase shift,

$$\mathbf{G}^{U(1)}(x)\mathscr{L}(l_L, e_R, B) \to \mathscr{L}(l_L', e_R', B').$$

Symbolically, we write the combination of the two

$$\mathbf{G}^{\{SU(2)_L^W \times U(1)\}}(x)\mathscr{L}_1(l_L, e_R, W, B) \to \\ \mathscr{L}_1(l_L', e_R', W', B').$$

At this point we add two further terms to the Lagrangian. Firstly, we add the Higgs wavefunctions with the cleverly chosen shape to their interaction energy. As this term must also be locally gauge invariant it must contain both gauge particles:

$$\mathscr{L}_2(\phi, B, W).$$

Secondly, we may allow the Higgs particles to interact with the leptons as we have yet to generate their mass

$$\mathscr{L}_3(l_L, e_R, \phi, B, W).$$

We now break the local gauge invariance of the total Lagrangian $\mathscr{L}_1 + \mathscr{L}_2 + \mathscr{L}_3$ under the combined $SU(2) \times U(1)$ group by redefining the Higgs wavefunction such that the state of zero wavefunction strength coincides with the minimum of the Higgs interaction energy we have introduced.

$$\phi \to \phi' - R \quad \mathscr{L}_2(\phi, B, W) \to \mathscr{L}_2(\phi', B', W'),$$
$$\mathscr{L}_3(l_L, e_R, \phi, B, W) \to \\ \mathscr{L}_3(l_L', e_R', \phi', B', W').$$

Once the mathematical smoke has cleared following this redefinition the following features emerge.

The masses of the gauge bosons are generated by the mixing of R with B and W in $\mathscr{L}_2$ and $\mathscr{L}_3$. However, as we have noted, the photon must remain massless whilst the W bosons become massive. We can see how this happens when we notice that neither the W-gauge particle of weak isospin nor the B-gauge particle of weak hypercharge can be identified with the electromagnetic gauge field of the photon. Just as we have defined the electric charge to be a mixture of weak isospin and weak hypercharge, the electromagnetic gauge particle generated by electric charge is a mixture of the wavefunctions of the weak isospin gauge particle W, and the weak hypercharge gauge particle B:

$$A = f_1(W^{0\prime}, B', \theta_W),$$

where the weak angle θ_W is the parameter which

Table 23.1. *The weak quantum numbers of the leptons*

	I^W	I_3^W	Y	Q
v_e	$\frac{1}{2}$	$\frac{1}{2}$	-1	0
$\dot{e}_L$	$\frac{1}{2}$	$-\frac{1}{2}$	-1	-1
e_R	0	0	-2	-1

adjusts the relative proportions of the two. The remainder of the W and B wavefunctions also mix together to produce another:

$$Z^0 = f_2\,(W^{0\prime}, B^\prime, \theta_W).$$

This is just the part of the weak interaction which has the same quantum numbers as the photon (i.e. zero electrical charge). It is the neutral current mentioned previously, and the uncharged gauge boson is now denoted Z^0 to signify its origin as a combination of the two fundamental gauge particle wavefunctions.

By careful choice of the form of the Higgs wavefunctions, it is possible to give a mass to the Z^0 mixture, whilst making sure that the A mixture remains massless. The charged components of the W-gauge particles suffer from none of these comcations and are simply given a mass by the sponspontaneous symmetry breaking. The values which emerge from the mathematics are:

$$M_{W\pm} = \frac{37}{\sin\theta_W}\ \text{GeV} \quad M_{Z^0} = \frac{74}{\sin 2\theta_W}\ \text{GeV}.$$

As the sine of an angle is always less than one, we can see that the model predicts the $W^\pm$ bosons to be at least 37 times the mass of the proton! This explains why the W bosons have been so difficult to detect experimentally. Not until very recently have our accelerators had enough energy to produce such massive particles.

The masses of the leptons are generated in $\mathscr{L}_3$ by the mixture of l_L, e_R and R. However, this exercise does not have the same predictive power as the mass generation of gauge bosons. Because we have incomplete knowledge of how the Higgs bosons might interact with the leptons, we are free to choose coefficients multiplying $\mathscr{L}_3$. So it is simple to ensure that the lepton mass terms are multiplied by the correct electron mass, and that the largely unknown quantity R is cancelled out.

23.3 Reprise

This, then, is the Glashow–Weinberg–Salam model. It has introduced the neutral current in a natural fashion, and discovery of the neutral current reactions in 1973 was a great boost to the acceptability of the model. The great cleverness of the model is to ensure the masslessness of the photon, whilst giving mass to the weak interaction $W^\pm$, Z^0 gauge bosons. This is achieved by introducing a suitable choice of Higgs wavefunctions. The masses generated are determined by the value of the one free parameter of the model, the weak angle. This must be measured, although there are some *ad hoc* predictions of its value ($\sin^2\theta_W \sim 0.25$). In addition to the neutral current, the model also requires the existence of a Higgs particle as a *bona fide* physical entity, although here some speculative theories can avoid this necessity.

23.4 An academic postscript – renormalisability

Although the model was essentially formulated as above in 1967 and 1968, it was not enthusiastically received until its renormalisability had been demonstrated. As we have mentioned, the local gauge invariance of the Lagrangian prior to spontaneous symmetry breaking suggested that it would be, but it remained to be shown that the symmetry breaking itself did not spoil this property.

't Hooft's proof of renormalisability essentially consisted of showing that the Feynman rules of the theory lead to mathematical expressions for the W-boson propagators, which avoid the problems associated with the use of massive spin-1 particles in perturbation theory, see Chapter 20. He showed that when the momentum flowing through a W-boson propagator is very large, then the mathematical expression of that propagator does not depend on the mass at all. As there is no mass dependence, there is no need for compensating momentum factors to ensure that the propagator is a dimensionless number. It is these extra momentum factors which lead to the divergences (infinite values) when summing over all the internal momentum configuration of a complicated Feynman diagram; without them, the answers are finite and thus the theory is renormalisable.

(The mechanism of the proof is roughly as follows. In order to write down the Feynman rules of a gauge theory and perform explicit calculations, it is necessary to break the gauge invariance and choose a particular gauge. This is just a convention and the physical answers obtained do not depend on the choice made. However, a choice must be made to be able to associate a unique expression with the propagator of the gauge particle. In theories with spontaneous symmetry breaking, the removal of the unwanted Goldstone boson is equivalent to a particular choice of gauge and, in the beginning, it seemed obvious to choose that gauge in which this unwanted particle did not exist. The result is that the term corresponding to an unphysical mixture of the Goldstone boson ϕ_1' and the gauge particle W which appears after spontaneous symmetry breaking becomes equal to zero. 'tHooft's method was to leave the choice of gauge open and retain this term, but to add to the Lagrangian yet another term which would cancel its effects. The effect of this is to produce a gauge particle propagator which does not depend on the gauge boson mass, but which contains factors describing the Goldstone bosons.

The factors in the gauge particle propagator exactly cancel the propagator representing a potentially real Goldstone boson, and so the unwanted particle never appears as a physical entity. Prior to 'tHooft's work, the choice of gauge was that in which the Goldstone boson was explicitly absent, but so too was its remedial effect on the W-boson propagator.)

24

Consequences of the model

24.1 Introduction

The Glashow–Weinberg–Salam model is now the undisputed theory of the weak (and electromagnetic) interactions. It has become established over the decade 1973–83 by a series of experiments which have confirmed the model's predictions. First, in 1973, the discovery of neutral currents provided a qualitatively new phenomenon, just as predicted by the model. Soon after, in the mid-1970s, new mesons were discovered which support the existence of a new quark type carrying the 'charm' quantum number – just as had been suggested by theorists attempting to describe the weak interactions of hadrons using the Glashow–Weinberg–Salam model. Next, in the late-1970s, various parity-violating effects were found to be in close agreement with the quantitative predictions of the model. Finally, and perhaps most impressively of all, between January and June 1983, positive evidence has been found for the existence of the $W^{\pm}$, Z^0 bosons at masses very close to those suggested by the model.

We now describe, in turn, each of these consequences of the model, except for the discovery of the $W^{\pm}$ bosons, which we shall come to in Part 9.

24.2 Neutral currents

In the discussion of neutrino–electron scattering in Chapter 17, we floated the possibility of the existence of neutral current processes by which, for example, an incoming neutrino may not have to turn into a charged lepton but could emerge with its identity unscathed. The emergence of the neutral gauge boson Z^0 from the Glashow–Weinberg–Salam model provides a natural explanation for such neutral current phenomena, and their detection was seen in the early-1970s as being an important test for the validity of the model. Reac-

tions which proceed by the neutral current are, for instance, elastic muon-neutrino–electron scattering or quasi-elastic muon-neutrino–proton scattering:

$$\nu_\mu + e^- \rightarrow \nu_\mu + e^-,$$

$$\nu_\mu + p \rightarrow \nu_\mu + p + \pi^0.$$

It was possible to check for the existence of these reactions only when neutrino beams were intense and energetic enough to permit detailed accelerator experiments to be performed. This became reality in

Fig. 24.1. (*a*) A neutral current reaction photographed in the Gargamelle bubble chamber at CERN, and its interpretation (*b*). (Photo courtesy CERN.)

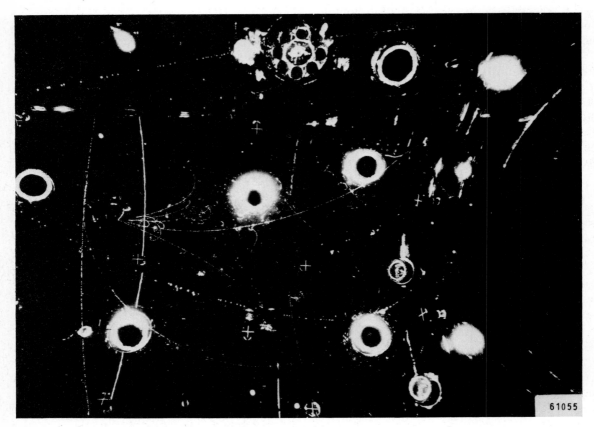

61055

$$\nu_\mu + p \longrightarrow \nu_\mu + p + \pi^+ + \pi^- + \pi^0$$
$$\qquad\qquad\qquad\qquad\;\; \downarrow \gamma$$

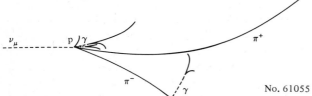

No. 61055

1973 when, much to everyone's surprise, neutral currents were found to be a significant effect, comparable in magnitude to the well-established charged currents. The process first seen was the second of the two mentioned above, see Figure 24.1, the magnitude of which, relative to its charged current version, was measured to be:

$$\frac{\sigma\ (\nu_\mu + p \to \nu_\mu + p + \pi^0)}{\sigma\ (\nu_\mu + p \to \mu^- + p + \pi^+)} = 0.51 \pm 0.25,$$

showing them to be effects of the same order. Since the first observation of this reaction, other neutral current processes have been observed as expected by the Glashow–Weinberg–Salam model, including the elastic muon-neutrino–electron scattering mentioned above.

All these reactions allow us to determine the value of the Weinberg angle which is the parameter which essentially fixes the relative mixing of the weak and electromagnetic interactions. The value which represents the average of the different experiments conducted to date is about

$$\sin^2 \theta_W = 0.20 \pm .03.$$

As we saw in Chapter 23, the masses of the intermediate $W^\pm$ bosons depend directly on this quantity which therefore predicts

$$M_{W^\pm} \approx 83 \quad \text{GeV} \qquad M_{Z^0} \approx 93 \quad \text{GeV}.$$

As we will see in Part 9, the most recent evidence from the CERN $p\bar{p}$ experiment provides spectacular confirmation of the $W^\pm$, Z^0 mass predictions.

24.3 The incorporation of hadrons–charm

It is necessary also for the Glashow–Weinberg–Salam model to describe the weak interactions of hadrons. In Chapter 18, we saw how these could be described using a weak interaction current written in terms of quarks. This form is the most convenient for investigating the consequences of the Glashow–Weinberg–Salam model. The charged current describing weak interactions where electric charge is changed is the same as before. It comprises a part which changes strangeness and a part which conserves strangeness, the relative importance of the two being regulated by the Cabbibo angle θ_C.

However, the Glashow–Weinberg–Salam model now requires an electrically neutral current, which, in the absence of further specifications, will

similarly consist of a part which changes the strangeness of the particles and a part which conserves it. An example of the strangeness-changing neutral current is given by the decay of the long-lived neutral kaon.

$$K_L^0 \to \mu^+ + \mu^-,$$

which is described in terms of quark interactions by the diagram (Figure 24.2(*a*)). Unfortunately, these processes, apparently allowed in the model, do not occur in the real world. Experimentally, the probability of this decay occurring is less than one part in one hundred million. So the model must be modified to ensure that these processes do not occur.

Several explanations were advanced in an attempt to cure this problem. The most successful, later to be rewarded by striking experimental evidence, is the 'charm' scheme proposed in 1970 by the international triumvirate of Glashow, Iliopoulos and Maiani, commonly known as GIM. The key assumption in the scheme is the existence of a new 'charmed' quark, c. This quark is to carry a new quantum number, charm, just as the strange quarks carry strangeness. All the other quarks, u, d, and s, are assigned zero charm. It is then possible to arrange that the unwanted strangeness-changing neutral current of Figure 24.2(*a*) is cancelled out by the corresponding diagram involving the charmed quark, shown in Figure 24.2(*b*).

In this way, it is possible to arrange for the disappearance of all the unwanted current in reactions such as

$$K^+ \to \pi^+ + \mu^+ + \mu^-,$$
$$K^+ \to \pi^+ + e^+ + e^-.$$

The consequences of the GIM scheme are enormous. The existence of this fourth quark flavour implies the existence of whole new families of charmed mesons and charmed baryons, rather like a repeat showing of all the strange particles. What is more, for the charmed quark diagram to cancel out the unwanted reactions effectively requires the mass of the charmed quark (and so particles) to be relatively small at about 1.5 GeV. So we should expect to see charmed particles copiously produced in our modern accelerators.

The charm scheme seemed to many physicists to be rather an 'expensive' way of solving the problem; having to introduce whole families of unknown particles just to alter the reaction rate of a small obscure class of reactions. However, this scepticism soon turned to amazement as the required particles tumbled out of the accelerators in the mid-1970s (see Chapter 35).

24.4 Parity-violating tests of the Glashow–Weinberg–Salam model

In addition to the qualitatively new phenomena described in the previous two sections, the Glashow–Weinberg–Salam model also tells us the magnitude of the various parity-violating effects due to the weak force. The most convincing evidence in this area is provided in polarised electron–deuteron scattering in which all the electrons can be collided with their spins pointing in a specified direction. The experiment measures the difference in the scattering cross-sections between left- and right-handedly polarised electrons off polarised deuteron targets. Not only is the magnitude predicted correctly, giving a difference between cross-sections of about 0.01%, but the way this difference changes with the energy of the collision is also explained.

Fig. 24.2. Strangeness-changing neutral currents which can proceed by the intermediate quark process in (*a*) can be cancelled out of the model by additional quark process involving the charmed quark in (*b*).

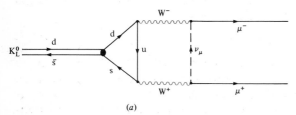

(*a*)

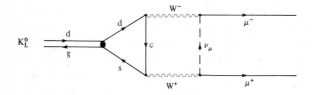

(*b*)

Another interesting parity-violating experiment is that concerning the interaction of light with matter at low energies. This provides a test of the Glashow–Weinberg–Salam model in a wholly new domain, and is thus that much stronger a test of its general worth. There are many variants of the experiment, but the basic idea is to shine a beam of polarised light (light whose electric field vector is aligned in a specific direction) through a vapour of metal atoms. The light is absorbed and reemitted by the various transitions of the atomic electrons and, because of the effects of the weak interaction between the atomic nucleus and the electrons, this leads to a very slight, but well-defined, rotation of the plane of polarisation in a given direction. Such a given rotation, of course, implies a distinction between clockwise and anticlockwise and so is indicative of parity violation. Because the effect is slight and because there are uncertainties about using the Glashow–Weinberg–Salam model in the atomic environment, the experiments and their interpretation are extremely hard work. But after initial doubts and discrepancies, there is now a convincing consensus of experimental results supporting the model. The Glashow–Weinberg–Salam model also predicts novel effects in electron–positron annihilations (see Chapter 34) and these also provide support. The model is now our standard working understanding of the electromagnetic and weak interactions (sometimes now referred to as the unified electroweak force) and in recognition of this, Glashow, Weinberg & Salam were awarded the Nobel prize in 1979.

Part 6
Deep inelastic scattering

25
Deep inelastic processes

25.1 Introduction

Among the most important experiments of the last 15 years have been those which use the known interactions of the leptons to probe the structure of the nucleons. Their importance lies in the fact that they provided the first dynamical evidence for the existence of quarks, as opposed to the static evidence of the success of the internal symmetry scheme $SU(3)$.

The name 'deep inelastic scattering' arises because the nucleon which is probed in the reaction nearly always disintegrates as a result. This is obvious from the momentum–wavelength relation for particle waves:

$$p\lambda = h. \tag{25.1}$$

The proton is approximately 10^{-15} m in diameter and so to resolve any structure within this requires that the probing particle wave has a smaller wavelength. The formula then gives the required momentum of the probe as being greater than 10 MeV/c, under the impact of which the target nucleon is likely to disintegrate.

Deep inelastic experiments divide into two classes depending on the nature of the probe used, which in turn dictates the force involved. In electroproduction, electrons or muons are scattered off the target nucleon and the force involved is electromagnetic. The leading process of the scattering is that of single-photon exchange, which is assumed to be a sufficiently good description of the interaction, see

Figure 25.1(*a*), although, in principle, more-complicated multi-photon processes may become significant as the energy of the collision becomes very large. The second class of experiment is called neutrinoproduction and, in this, neutrinos are scattered off the target nucleon by the weak nuclear force. The leading process is that of single-W-boson exchange, other more complicated processes being insignificant. Both charged and neutral currents may contribute, see Figure 25.1(*b*), (*c*), but in practice it is the better-understood charged currents which are used in experiments. Indeed, this was necessarily the case in the early experiments, 1967–73, as the neutral currents had not then been discovered.

The main measurement of the experiments is the variation of the cross-section (the effective target area of the nucleon) with the energy lost by the lepton during the collision and with the angle through which the incident lepton is scattered. The energy lost by the lepton v is simply the difference between its incident and final energy:

$$v = E_i - E_f. \tag{25.2}$$

Fig. 25.1. (*a*) Electroproduction via photon exchange. (*b*) Charged current neutrino-production via $W^\pm$ exchange. (*c*) Neutral current neutrinoproduction via Z^0 exchange.

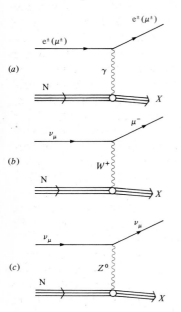

The angle through which the lepton is scattered is related to the square of the momentum transferred by the photon q^2 from the lepton to the nucleon by the formula:

$$q^2 = 2E_i E_f (1 - \cos \theta). \tag{25.3}$$

These are the two main observables in deep inelastic scattering, which connect the data from experiments with our theoretical picture of the nucleon interior.

25.2 Two key ideas

Two ideas in particular played an important role in the development of the experiments and in our understanding of them. The two ideas, both put forward in 1969, are those of the parton model and of scaling.

The parton model was first put forward by Richard Feynman and is simply a formal statement of the notion that the nucleon is made up of smaller constituents; the partons. No initial assumptions about the partons are necessary as it is the purpose of the experiments to determine their nature. But obviously we have at the back of our minds the identification of the partons with the quarks of $SU(3)$. However, we should not jump to the conclusion that *only* the familiar quarks will be sufficient to describe the composition of the nucleon. For instance, in addition to the proton's three quarks which are required by the internal symmetry scheme (the so-called 'valence' quarks), it may be possible for other quark–antiquark pairs to emerge briefly from the vacuum by borrowing energy according to Heisenberg's uncertainty principle. These 'sea' quarks may then form an additional material presence within the nucleon and provide a mechanism for the existence, albeit transient, of antimatter inside a 'matter' particle. Because they emerge in quark–antiquark pairs, the sea quarks will have no net effect on the quantum numbers of the nucleon generated by the valence quarks. In addition to the quarks, we may be alert to the fact that quanta of the interquark force field may also be present inside the proton. Just as electrons interact by the exchange of photons, the quanta of the electromagnetic force, so quarks may interact by the exchange of quanta of their force field. These hypothetical quanta have been called, rather simplistically, gluons, because they glue the quarks together.

Scaling is the name given to a phenomenon of the cross-section first predicted by the Stanford physicist James Bjorken. Stated simply, the prediction is that when the momentum carried by the probe becomes very large, then the dependence of the cross-section on parameters such as the energy v and momentum squared q^2, transferred by the photon, becomes very simple. In the parton model, the onset of this simple scattering behaviour has a straightforward interpretation. The complicated scattering of the probe off a nucleon of finite spatial extent has been replaced by the scattering of the probe off a point-like parton. The photon ceases to scatter off the nucleon *as a coherent object* and, instead, scatters off the individual point-like partons *incoherently*. We should expect this sort of behaviour to manifest itself when the wavelength of the probe is much less than the nucleon diameter, implying a probe momentum above about 1 GeV.

Observation of this scaling behaviour in 1969 immediately lent support to the parton model of the nucleon, although, we will see, the initial discovery was somewhat fortuitous. To understand the concepts of scaling and the parton model further, we must take a more detailed look at the processes involved.

26

Electron–nucleon scattering

26.1 Introduction

Assuming that the electromagnetic interaction between the electron and the nucleon is dominated by the single-photon exchange mechanism, then the mathematics used to describe the reaction becomes relatively simple. To check the experimental observations, we want to derive a formula to explain how the cross-section varies with the energy v and momentum transfer squared q^2 of the intermediate photon. The formula is made up of factors associated with the different parts of the diagram in Figure 25.1(*a*). It consists of a factor describing the progress of the electron through the reaction (the lepton current), a factor describing the propagation of the virtual photon as a function of v and q^2, and a factor describing the flow of the nucleon in the reaction including the complicated disintegration process (the hadron current). The factors describing the electron and the photon are well known from QED and present us with no problems. But the factor describing the hadron current is a complicated unknown describing the evolution of nucleon structure during the reaction. This unknown can be characterised by a number of 'structure functions' of which we assume no prior knowledge and which are to be determined by the deep inelastic experiments, see Figure 26.1.

The format for the structure functions is discovered by writing down the most-general possible combinations of all the momenta appearing in the reaction and then simplifying the result using general theoretical principles such as parity and time-reversal invariance.

The two separate functions of q^2 and v resulting, $F_{1,2}(q^2, v)$, correspond to the scattering of the two possible polarisation states of the virtual photon exchanged; longitudinal and transverse. The longitudinal polarisation state exists only because of the virtual nature of the exchanged photon (because it temporarily has a mass). On the mass shell when the virtual photon becomes real (massless), then the longitudinal polarization state and its associated structure function disappears. The separate behaviour of the two structure functions can be determined from experiments because they are multiplied by coefficients involving different functions of the electron scattering angle. By observing the reaction at different values of this angle, the two behaviours can be separated out.

26.2 The scaling hypothesis

The scaling hypothesis mentioned previously is to do with these structure functions. It is important to realise that they are just numbers and have no physical dimension. The cross-section is usually given in units of area which are provided by the simple Rutherford scattering formula for elastic scattering. This has deep implications for the behaviour of the structure functions. If they are to have any dependence on physically dimensional quantities such as the energy v or momentum squared q^2 involved in the reaction, then these factors must have their physical dimensionality cancelled out to give structure functions in terms of pure numbers.

In low-energy elastic scattering ($v = 0$), the photon effectively perceives the nucleon as an extended object and the structure function essen-

tially describes the spatial distribution of electrical charge on the nucleon. This leads to a dependence of the structure function on the momentum of the photon – but the dimensionality of the momentum in the structure function is cancelled out by factors of the nucleon mass:

$$\frac{d\sigma}{dq^2} = \frac{4\pi\alpha^2}{q^4} \cdot F\left(\frac{q^2}{M_N^2}\right), \tag{26.1}$$

cross-section = unit of area × pure number.

To signify this cancellation we say that the nucleon mass sets the 'scale' of reaction. It provides a scale against which the effect of the photon momentum can be measured.

In very high-energy deep inelastic scattering ($q^2, v \rightarrow \infty$), the photon has resolved down to such an extent that the existence of the complete nucleon is really irrelevant to the reaction, the photon interacts with only a small part of the nucleon and does so independently of the rest of it. This means that there is no justification for using the nucleon mass to determine the scale of the reaction. In fact, there is no justification for using any known mass or any other physically dimensional quantity to determine the scale of the deep inelastic regime. James Bjorken grasped the consequences of this abstract argument: if the structure functions are to reflect the dependence of the cross-section on the shape of the nucleon as seen by a photon of very high q^2 and v, and if there exists no mass scale to cancel out the physical dimensions of these quantities, then the structure functions can only depend on some dimensionless ratio of the two. Choosing such a ratio as x

$$x = \frac{q^2}{2M_N v}, \tag{26.2}$$

then the scaling hypothesis is that the structure functions can depend only on it, and not on either or both of the quantities involved separately.

Fig. 26.1. The formula describing the differential cross-section for electron–nucleon scattering with respect to the momentum transfer squared q^2 and the energy lost by the electron v. The structure functions F_1 and F_2 essentially describe the shape of the nucleon target.

$$F_{1,2}^{eN}(q^2, v) \xrightarrow[q^2, v \rightarrow \infty]{} F_{1,2}^{eN}(x). \tag{26.3}$$

$$\frac{d^2\sigma}{dq^2 dv} = \frac{4\pi\alpha^2}{q^4} \frac{E_f}{E_i M_p} \left[\frac{M_p}{v} F_2(q^2, v) \cos^2\frac{\theta}{2} + 2F_1(q^2, v) \sin^2\frac{\theta}{2} \right]$$

The scaling hypothesis becomes rather more accessible when it is combined with the parton model in which the nucleon is regarded as a simple collection of point-like constituents. A point has no mass and no dimension, and we are considering the scattering of a photon carrying infinitely high momentum (one which has a vanishingly small wavelength). In this situation, there are simply no physically significant quantities which are relevant to set the scale of the reaction. So quantities such as the energy and momentum transfer squared in the reaction can only enter into its description in the form of pure numbers, which in turn implies a dimensionless ratio of the two.

Bjorken's choice of the 'scaling' variable x has a very significant interpretation. It turns out to be the fraction of the momentum of the nucleon carried by the parton which is struck by the photon. So the structure functions, which depend only on x, effectively measure the way in which the nucleon momentum is distributed amongst its constituent partons.

Figure 26.2(*a*) shows how the structure function $F_2^{ep}(x)$ varies with x, as measured in early experiments at the Stanford Linear Accelerator Center. As can be seen, the shape implies that the majority of collisions occur with partons carrying a relatively small fraction of the nucleon momentum. Figure 26.2(*b*) shows a test of the scaling hypothesis that the structure function depends only on x, and not on q^2 (or v) separately. At a given value of x, the structure function is found to be almost constant over a q^2 range from 1 to 8 GeV². Data such as this became available in the early 1970s. The apparent validity of the scaling hypothesis and the plausibility of its connection with the existence of point-like constituents within the nucleon led to a more detailed investigation of the structure functions to establish more information about the mysterious partons.

26.3 Exploring the structure functions

One straightforward exercise is to compare the formula for electron–proton scattering with formulae derived from QED describing the electromagnetic interactions of electrons with other simple, electrically charged particles whose properties, such as spin, may be assumed. In this way it is possible to derive relationships between the struc-

ture functions depending on the similarities between the properties of the partons and those of the hypothetical particles assumed in the QED formulae. For instance, by comparing the formula of Figure 26.1 with the QED formula describing the scattering of an electron with an electrically charged particle of spin $\frac{1}{2}$, it is possible to derive the relationship between the structure functions.

$$2xF_1(x) = F_2(x). \qquad (26.4)$$

Thus if, experimentally, the ratio $2xF_1/F_2$ is found to be equal to one, this may be interpreted as evidence for the partons having spin $\frac{1}{2}$. Using similar formulae it is possible to show that if the ratio is observed to be zero, then this provides evidence for spin-0 patrons. As can be seen from Figure 26.3, the evidence is firmly in favour of spin-$\frac{1}{2}$ partons.

Fig. 26.2. Early data on scaling. In (*a*), $F_2(q^2,v)$ is a universal function of x for a range of values of q^2 (many experimental points are contained in the shaded area). In (*b*), $F_2(q^2,v)$ demonstrates approximate constancy over the range of q_2 measured in the early experiments.

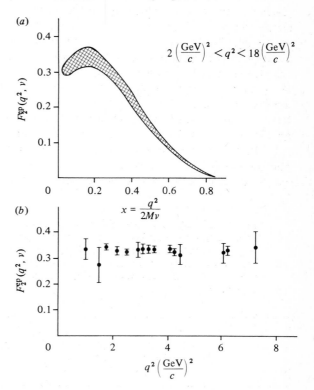

Another lesson to be learned by comparing the deep inelastic formula with the simpler formula from QED is that the structure functions essentially measure the distribution of electric charge within the nucleon. In low-energy, non-relativistic physics, it is acceptable to express this distribution of charge over the spatial extent of the proton. But in high-energy physics this is better expressed in terms of the conjugate description of how the nucleon momentum is distributed between the various electric charges of the partons. If we say that the ith type of parton with charge Q_i has probability $f_i(x)$ of carrying a fraction x of the nucleon momentum, then it is possible to relate the overall structure functions of the nucleon to these individual parton momentum distributions,

$$F_1^{eN}(x) = \sum_{i \text{ partons}} f_i(x)Q_i^2,$$

and

$$F_2^{eN}(x) = x \sum_{i \text{ partons}} f_i(x)Q_i^2.$$

(26.5)

Obviously, if we integrate the structure functions over the fractional momentum carried by the partons, we should then expect to provide some measure of the total charge carried by the partons. In fact the most convenient relationship involves the sum of the squares of the parton charges:

$$\int_0^1 \frac{F_2(x)}{x}\,\mathrm{d}x = \sum_i Q_i^2.$$

(26.6)

So, by investigation of relationships involving the structure functions, it is possible to find evidence for the spin of the partons and their charge assignments, which we will investigate further in Chapter 28.

Fig. 26.3. The ratio of the structure functions provides a test of the spin-$\frac{1}{2}$ assignment to the partons.

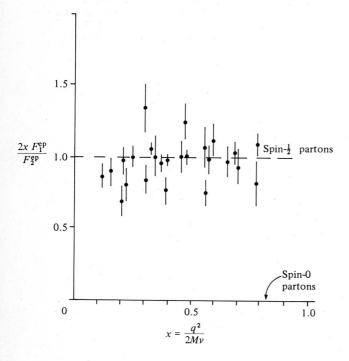

27

The deep inelastic microscope

27.1 Introduction

The physics behind the approach to scaling can be appreciated intuitively by regarding deep inelastic scattering as an extension of the ordinary microscope, whose successor the experiments quite literally are. The distance to which an ordinary microscope can resolve depends ultimately on the wavelength of the light scattered from the object under view. The shorter the wavelength, the smaller the distances which can be resolved. The high-energy photon exchanged between the electron and the nucleon in deep inelastic experiments is simply the logical development of the microscope technique. The origin of the scaling phenomena becomes clear by regarding a succession of snap-shots of the virtual photon–nucleon collisions:

When the momentum carried by the photon is low, its wavelength is relatively long compared to the dimensions of the nucleon. It will not be able to resolve any structure and will effectively see the nucleon as a point. In this case, structure functions are irrelevant, being represented simply by the nucleon charge in the Rutherford formula for the scattering of two point charges, see Figure 27.1(a).

With higher momentum, the photon will have a wavelength comparable to that of the nucleon. The photon will begin to resolve the finite spatial extent of the nucleon and the structure functions will depend in a non-trivial fashion on the momentum carried by the photon, modifying the Rutherford scattering cross-section, Figure 27.1(b).

Eventually, with high-momentum transfer the photon will have a very short wavelength and may resolve the internal structure of the nucleon. Scattering off point-like constituents within the nucleon is indicated when the structure function assumes a simple dependence on some dimensionless 'scaling variable'. In fact, under certain circumstances the structure functions are simply replaced by a constant equal to the sum of the squares of the charges of the constituents, which multiplies the simple Rutherford cross-section, Figure 27.1(c). This reversion to the simplicity of point-like scattering after a relatively more complicated transitional phase is taken to indicate that we have broken through to a new, more basic level of matter within the nucleon.

27.2 Free quarks and strong forces

An essential consequence of the validity of the scaling hypothesis, in which the deep inelastic probe scatters *incoherently* off the individual partons, is that the partons are essentialy free from mutual interactions over the space–time distances of the

Fig. 27.1. An illustration of the approach to scaling as the virtual photon wavelength becomes much smaller than the nucleon diameter. The wavelengths become progressively smaller in (a), (b) and (c).

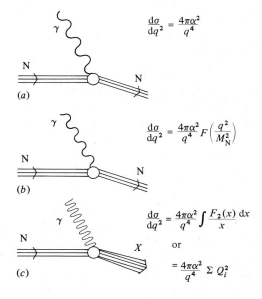

$$\frac{d\sigma}{dq^2} = \frac{4\pi\alpha^2}{q^4}$$

(a)

$$\frac{d\sigma}{dq^2} = \frac{4\pi\alpha^2}{q^4} F\left(\frac{q^2}{M_N^2}\right)$$

(b)

$$\frac{d\sigma}{dq^2} = \frac{4\pi\alpha^2}{q^4} \int \frac{F_2(x)\, dx}{x}$$

or

$$= \frac{4\pi\alpha^2}{q^4} \Sigma\, Q_i^2$$

(c)

probe–parton interaction. This has important consequences for the nature of interparton forces.

To understand this more fully, we can associate one interaction time with the duration of the probe–parton interaction τ_1 and another time to characterise the duration of interquark forces τ_2. Obviously, for the nucleon to have any sort of collective identity, the nucleon must exist for $\tau_{\text{life}} > \tau_2$, so that the partons can become aware of each others' presence. As a rough guide we may estimate the interaction times by the following prescriptions:

$$\tau_1 = \frac{\text{wavelength of probe}}{\text{speed of light}} \approx \frac{1}{v},$$

$$\tau_2 = \frac{\text{interquark distance}}{\text{speed of light}} < \tau_{\text{life}}.$$

If $\tau_2 \lesssim \tau_1$, then the interparton forces will have transmitted the effects of the probe collision to all the partons inside the nucleon within the probe's interaction time. In this case, the probe will not be scattering off the individual partons but off the entire nucleon instead. However, in deep inelastic scattering in which the probe wavelength is very small, $\tau_1 \ll \tau_2$ and therefore the probe interaction is completed well before the interparton forces have had time to relay the event to the rest of the nucleon. So the probe–parton interaction occurs well within the lifetime of the nucleon.

For a short time, subsequently, the nucleon exists in an uncomfortable state: one of its partons has been struck hard and flies off with the high momentum imparted by the probe, but the other partons know nothing of this and continue to exist in a quiescent state, see Figure 27.2. This situation cannot last long, otherwise the struck parton would eventually appear as an isolated particle separate from the rest. This has never been observed. Instead, it is thought that final-state interactions come into effect between the struck parton and the others, turning the energy of the collision into the observed hadrons, Figure 27.2(c), (d).

An important feature of the parton model is the assumption that the cross-section for deep inelastic scattering can be calculated simply by summing over the individual probe–parton interactions and that the complicated final-state interactions become important only over longer space–

time distances (i.e. greater spatial separations and longer interaction times).

Thinking particularly in terms of the strong interactions between quarks, the overall picture emerging from the success of the parton model is that of a force whose strength varies with distance.

Fig. 27.2. When a deep inelastic probe strikes a parton (a) and (b), it flies off with large momentum (c) until some confining mechanism pulls extra parton–antiparton pairs from the vacuum, creating new particles (d).

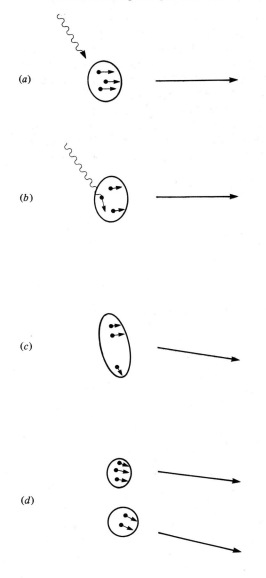

At the short distances probed by the deep inelastic reactions (say 10^{-17} m) the interquark force is weak and the quarks are essentially free. But as the distance between two quarks grows to the nucleon diameter (10^{-15} m), the force also grows, confining the quarks, perhaps permanently, within the observed hadrons. Eventually, any energy expended in trying to separate the quarks will be sufficient to pull a new quark–antiquark pair from the vacuum, so allowing the production of a new hadron, but preventing the emergence of the individual quarks.

The tendency of the interquark forces to become weaker at small distances is known as asymptotic freedom. As we shall see in Part 7, the discovery that this property can be explained naturally in non-Abelian gauge theory was a significant breakthrough in our attempt to understand quark dynamics. The other tendency of the interquark forces, to become increasingly strong as the quarks are separated, is known as confinement. As discussed also in Part 7, a satisfactory description of this phenomenon remains an outstanding problem.

28

Neutrino–nucleon scattering

28.1 Introduction

Just as in electron– or muon–nucleon scattering the exchanged photon acts as a probe of the electromagnetic structure of the nucleon, so in neutrino (or antineutrino)–nucleon scattering, the exchanged W boson probes the distribution of 'weak charges' within the nucleon. For this purpose, the two most important processes are the charged current inclusive reactions, see Figure 25.1 An important feature of these reactions is that they are able to distinguish between the partons and the antipartons of the target nucleon. This is because the space–time structure of the weak interaction ensures that target partons of differing helicities are affected differently. In the relativistic limit, in which the rest mass of a particle is regarded as being negligible, then parton and antiparton helicities are opposite, so they will interact with the W-boson probe differently. Also, because the W-boson probe is electrically charged, the target parton must be able to absorb the charge. As we shall see, this rules out the participation of some types of parton, making the weak interaction a more selective probe of the nucleon's interior than the indiscriminate photon.

28.2 Neutrino experiments

Although these weak interaction experiments are theoretically more illuminating than their corresponding electromagnetic counterparts, the practical difficulties of dealing with neutrinos tend to spoil their potential.

The electrically neutral, weakly interacting neutrinos cannot be directed by electric and magnetic fields, as can electrons; and building a usable neutrino beam is a complicated process. Firstly, a primary beam of protons is accelerated to a high energy and is made to collide with a stationary target such as a piece of iron. From these collisions, a host of secondary particles, mainly mesons, will emerge in the general direction of the incident proton beam, although with somewhat lesser energies. These secondary mesons can then decay into neutrinos or antineutrinos and various other particles by decays such as:

$$\pi^{\pm} \rightarrow \mu^{\pm} + \nu_{\mu} \ (\text{or} \ \bar{\nu}_{\mu}).$$

Because the muon-decay mode of the mesons is generally the most common, it is mainly muon-type neutrinos which make up the beam. Finally, the neutrinos are isolated by guiding the secondary beam through a barrier of, perhaps, 0·5 km of earth. Only the weakly interacting neutrinos can pass through this amount of matter and so the beam emerging from the far side of the barrier is a pure neutrino beam with a typical intensity of about 10^9 particles per second. Unfortunately, the initial proton–target collisions and the subsequent decays of the secondary mesons mean that the resulting energy of the neutrino beam is rather uncertain and often must be inferred by adding up the energies of the products of the neutrino–nucleon interactions under study.

To obtain a reasonable rate of interactions, the neutrino beam must be passed through a very massive target. For instance, the Gargamelle bubble chamber at CERN contains about ten tons of some heavy liquid such as freon to ensure a satisfactory rate of reactions. Because the beam consists of both neutrinos and antineutrinos, both sorts of reaction will occur during the same experiment. The two can be distinguished by observing the charge of the outgoing muon in any particular reaction.

28.3 The cross-section

For $\nu_{\mu}N$ scattering, this may be written down in a fashion similar to that used for $e^{\pm}N$ scattering, by combining various factors describing the different sub-processes which go to make up the collision. Referring back to figure 25.1, we can see that these factors must include one to describe the transformation of the incoming neutrino into a muon by emitting the W boson (the lepton current); one describing the propagation of the W boson; and one describing the disintegration of the target nucleon under the impact of the W boson (the hadron current). Also analogous to the case of $e^{\pm}N$ scattering is that the behaviour of the factors is well known apart from that describing the hadron current. The lepton current and the propagation of the W boson is well known from the gauge theory of the weak interactions – which is well approximated by the simpler Fermi theory at all practicable energies. But, as before, the unknown form of the hadron current must be characterised by a number of structure functions whose nature it is the job of the experiments to discover, see Figure 28.1.

The format of the structure functions is found as before by writing down all the possible combinations of momenta involved in the reaction and then appealing to general principles to simplify the result. In contrast to the electromagnetic force, the weak force does not respect parity invariance and so this simplifying influence cannot be used in neu-

Fig. 28.1. The formula describing the differential cross-section for (anti)neutrino–nucleon scattering with respect to momentum transfer squared q^2 and energy lost by the neutrino ν. Three structure functions F_1^W, F_2^W and F_3^W (functions of q^2 and ν in general) are needed to describe the shape of the nucleon target (i.e. the way in which weak charge is distributed over the nucleon momentum).

$$\frac{d^2\sigma}{dq^2 d\nu} = \frac{G_F^2}{2\pi} \frac{E_{\mu}}{E_{\nu(\bar{\nu})}} \left[2F_1^W(q^2, \nu) \sin^2 \frac{\theta_{\mu}}{2} + F_2^W(q^2, \nu) \cos^2 \frac{\theta_{\mu}}{2} \pm F_3^W(q^2, \nu) \frac{(E_{\mu} + E_{\nu(\bar{\nu})})}{M_N} \sin^2 \frac{\theta_{\mu}}{2} \right]$$

trino–nucleon scattering. Because of this, a third weak structure function is introduced (F_3^W) which enters with a different sign in the formula depending on whether neutrinos or antineutrinos are being scattered. This is the manifestation of the effect mentioned earlier by which the parity-violating weak interaction will distinguish between matter and antimatter involved in the reactions as a result of helicity effects. In general, the structure functions all depend on q^2 and v separately. It is interesting to note that it is *only* through the structure functions that these quantities enter the description of the reactions at all.

28.4 The scaling hypothesis

Although first described in connection with the electromagnetic interactions, this hypothesis is equally valid for the weak force. In this case, all the dimensionality of the cross-section is contained within the Fermi coupling constant G_F (remember that this was the trouble which provided one of the major motivations for the development of weak interaction field theory). As a result, the structure functions must be pure numbers. In the absence of any 'scaling factor' to cancel out the dimensionality of q^2 and v, the structure functions $F_{1,2,3}^W$ cannot depend on them as individual quantities, but only on some dimensionless ratio of the two:

$$F_{1,2,3}^W (q^2, v) \xrightarrow[q^2, v \to \infty]{} F_{1,2,3}^W (x), \qquad (28.1)$$

where the ratio x is the same as before. The structure functions can be measured directly as in $e^\pm N$ scattering and the scaling hypothesis tested. The general shape of the weak structure functions is much the same as that of the electromagnetic example illustrated in Figure 26.2, but because the parameters of the neutrino beam are that much more uncertain than of the electron or muon beam, the experimental errors are much larger, thereby providing a weaker test of the scaling hypothesis.

However, the scaling hypothesis does predict that a far more obvious characteristic will hold in neutrino–nucleon scattering. As mentioned earlier, the cross-section does not depend on q^2 or v apart from through the structure functions. If, then, this dependence is removed by the scaling hypothesis, it means that the cross-section will not depend on these quantities at all. In this case, it is possible to

integrate the formula for the cross-section over all possible values of q^2 and v in a very simple fashion to obtain the total cross-section for neutrino or antineutrino–nucleon scattering.

$$\sigma^{v(\bar{v})N} = \int \frac{d^2\sigma}{dq^2 dv} \, dq^2 dv \propto \frac{G_F^2 M E_{v(\bar{v})}}{\pi}. \qquad (28.2)$$

The resulting scaling prediction for neutrino–nucleon scattering is that the total cross-section should rise linearly with the energy of the incident neutrino. The slope of the rise is given by constants which will be different for neutrino or antineutrino reactions, because of the different signs in front of F_3 in the formula of Figure 28.1. Experimental measurements of the total cross-sections are consistent with the linear rise with energy predicted by the scaling hypothesis and its interpretation in terms of point-like partons carrying both electric and weak charges, see Figure 28.2.

The difference in the slopes of the energy dependencies of v and $\bar{v}$ scattering is measured to be about a factor of 3. This factor can be easily

Fig. 28.2. Total neutrino–nucleon and antineutrino–nucleon cross-sections plotted against energy support the scaling hypothesis by exhibiting a linear increase.

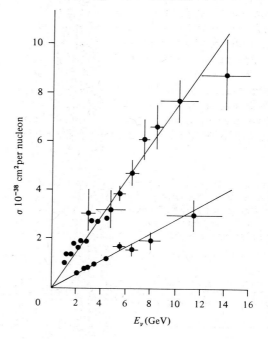

understood in terms of the underlying neutrino–parton interactions. Because they are massless, neutrinos exist only with left-handed spin and antineutrinos only with right-handed spin. Also, inasmuch as the mass of the partons may be neglected (i.e. in the relativistic limit) they too are solely left-handed (assuming spin-$\frac{1}{2}$ partons from the evidence of Figure 26.3). Because partons predominate over antipartons in the nucleon, the neutrino– and antineutrino–parton collisions can be distinguished by the way in which these spins add up. If the two colliding spins cancel each other out, as in the case of neutrino–parton scattering, then no restrictions are placed on the angles of emergence of the outgoing particles, see Figure 28.3(*a*). If, on the other hand, the two colliding spins add up, as in the case of antineutrino–parton scattering, then the existence of non-zero angular momentum in the system restricts the permitted angles of emergence of the outgoing particles. This means that the

cross-section of antineutrino–parton scattering is reduced relative to that of neutrino–parton scattering. This is because the integration over q^2 to obtain the total cross-section is equivalent to summing over all possible angles of emergence (bearing in mind the definition (25.3)) which are more restricted when the angular momentum is non-zero. The mathematics predicts a factor of three between νN and $\bar{\nu}$N scattering just as observed.

Neutrino–nucleon scattering provides an independent test of the scaling hypothesis and of the parton model. What is now possible is the comparison of muon–nucleon and neutrino–nucleon scattering to establish that the electromagnetic and weak interactions 'see' the same partons.

Bearing in mind that we believe the electromagnetic and weak interactions to be just the different manifestations of the same 'electroweak' force, we certainly expect that this should be the case. Also, we should like to compare the properties of the partons with those of the quarks of $SU(3)$.

Fig. 28.3. (*a*) Neutrino–parton scattering: spins cancel. There are no restrictions on the directions of the emerging particles. (*b*) Antineutrino–parton scattering: spins add. Restrictions limit the angles of emergence of the outgoing particles.

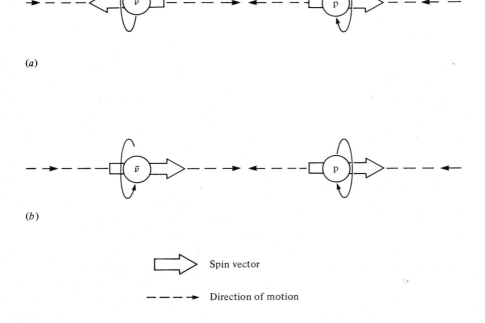

(*a*)

(*b*)

⇨　Spin vector

- - - →　Direction of motion

29

The quark model of the structure functions

29.1 Introduction

To be more specific about the content of the nucleon (i.e. about what makes up the structure functions), the approach adopted is to assume that the partons have the same properties as the quarks and then to work out the consequences for electron–nucleon and neutrino–nucleon scattering independently. Following this, it is possible to compare the results obtained to see if they are compatible. So we examine, in turn, the electron's eye view and the neutrino's eye view of the structure functions.

29.2 Electromagnetic structure functions

We have mentioned earlier the interpretation of the structure functions as the distribution of the squares of the parton charges within the nucleon, according to the fraction momentum x carried by the parton, given by formula (26.5)

$$\left.\begin{aligned} F_1^{eN}(x) &= \sum_i^{Nq} f_i(x)\, Q_i^2, \\ F_2^{eN}(x) &= x \sum_i^{Nq} f_i(x)\, Q_i^2. \end{aligned}\right\} \quad (26.5)$$

In the simple four-flavour quark model, the quarks and their charges are:

$$u\left(\frac{2}{3}e\right),\ d\left(-\frac{1}{3}e\right),\ s\left(-\frac{1}{3}e\right),\ c\left(\frac{2}{3}e\right),$$

where c denotes the fourth charmed quark of the GIM scheme with two-thirds of the charge on the electron. Assuming these values, we can write out the explicit quark content of the structure functions for both the proton and the neutron. In doing so, we must allow for the presence of both quarks and antiquarks from the vacuum sea. This gives, for the proton:

$$\begin{aligned} F_1^{ep}(x) = \Bigg[&\left(\frac{2}{3}\right)^2 (f_u(x)+f_{\bar u}(x)) + \\ &\left(\frac{1}{3}\right)^2 (f_d(x)+f_{\bar d}(x)) + \\ &\left(\frac{1}{3}\right)^2 (f_s(x)+f_{\bar s}(x)) + \\ &\left(\frac{2}{3}\right)^2 (f_c(x)+f_{\bar c}(x)) \Bigg]. \end{aligned} \quad (29.1)$$

The expression for the second structure function $F_2^{ep}(x)$ is just the same as above, only multiplied by x, and the corresponding expressions for the neutron structure functions, $F_1^{en}(x)$ and $F_2^{en}(x)$, can be obtained by interchanging $f_u(x) \leftrightarrow f_d(x)$ in the above expressions, as the distribution of up quarks inside the proton is equivalent to the distribution of down quarks inside the neutron.

The total fractional momentum carried by any particular sort of quark can be obtained simply by integrating over its momentum distribution. For instance, the total share of the momentum carried by the up quarks and antiquarks in the proton is given by:

$$P_u = \int_0^1 x(f_u(x)+f_{\bar u}(x))\, dx. \quad (29.2)$$

Similar expressions will obtain for other varieties of quark. The integrals involved are all contained within the integrals over the total structure functions which are measured in the experiments as the area under the distribution of Figure 26.2(*a*). This quantity gives a linear combination of the momentum shares of all the possible constituent quarks.

$$\left.\begin{aligned} &\int dx\, F_2^{ep}(x) = \\ &\quad \left(\frac{4}{9}P_u + \frac{1}{9}P_d + \frac{1}{9}P_s + \frac{4}{9}P_c\right) = \\ &\quad .18\ (\text{experiment}), \\[2mm] &\int dx\, F_2^{en}(x) = \\ &\quad \left(\frac{1}{9}P_u + \frac{4}{9}P_d + \frac{1}{9}P_s + \frac{4}{9}P_c\right) = \\ &\quad .12\ (\text{experiment}). \end{aligned}\right\} \quad (29.3)$$

In the above formula, it is fair to assume that the total fraction of the proton's momentum carried by the strange and charmed quarks is negligible. This assumption leaves us with two equations and two unknowns which may be solved to give:

$$P_u = .36, \quad P_d = .18. \tag{29.4}$$

Thus the total fractional momentum carried by the up quarks is measured to be twice that carried by the down quark, which supports the quark model's picture of the proton as (uud).

    However, the measurements also indicate that the total fractional momentum carried by the quarks is only one-half of the total proton momentum. The interpretation of this is that the other half is carried by neutral gluons which are the quanta of the strong nuclear force between the quarks. Because these gluons are electrically neutral, they do not experience the electromagnetic force and so show up only as missing momentum in the overall accounting for the proton.

29.3 Weak interaction structure functions

    The last section showed that the quark model can lend a great deal more detail to our picture of the structure functions. But the picture cannot be filled in completely because the photon transferred in electron–nucleon scattering differentiates between the quarks only by virtue of their electrical charge. To make further distinctions it is necessary to use the W-boson probe of the weak interaction. It is straightforward to establish which of the W-boson–quark interactions are possible, and which of those possible are dominant.

    Because of lepton-number conservation in charged current reactions, the neutrino must turn into a negatively charged muon and emit a positively charged W boson, and the antineutrino must turn into a positively charged muon emitting a negatively charged W boson. In principle, the W^+ boson can collide with a down quark, thereby changing it to an up, or it may collide with a strange quark, also turning it into an up. But the W^+ boson cannot be absorbed by an up quark, as this would result in a quark of charge $\frac{5}{3}e$ which does not exist. The possible reactions can be simplified further because the weak interactions which change strangeness, such as

$$W^+ + s\left(-\frac{1}{3}e\right) \rightarrow u\left(\frac{2}{3}e\right),$$

are much smaller than the strangeness-conserving weak interactions. This is just the Cabbibo hypothesis mentioned in Chapter 18. Because of it, we may ignore all the strangeness-changing reactions which may occur in principle. In addition to the interactions with quarks mentioned so far, the W bosons can also interact with the antiquarks from the vacuum sea. Combining all these considerations, we can summarise all the significant neutrino–quark interactions which are possible in neutrino–nucleon scattering:

(1)
$$\nu_\mu + d\left(-\frac{1}{3}e\right) \rightarrow \mu^- + u\left(\frac{2}{3}e\right);$$

(2)
$$\bar{\nu}_\mu + u\left(\frac{2}{3}e\right) \rightarrow \mu^+ + d\left(-\frac{1}{3}e\right);$$

(3)
$$\bar{\nu}_\mu + \bar{d}\left(\frac{1}{3}e\right) \rightarrow \mu^+ + \bar{u}\left(-\frac{2}{3}e\right);$$

(4)
$$\nu_\mu + \bar{u}\left(-\frac{2}{3}e\right) \rightarrow \mu^- + \bar{d}\left(\frac{1}{3}e\right).$$

$$\tag{29.5}$$

We may now proceed to write the cross-section for, say, neutrino–proton scattering as a sum of the neutrino–quark cross-sections involved.

    In doing this, it is usual to express the differential cross-section, not as before expressed as varying with the momentum squared q^2 and energy v carried by the intermediate W boson, but instead expressed as varying with the momentum fraction of the target carried by struck quark x and the fraction of the incident neutrino energy carried across by the W boson y. Mathematically, this requires us to make the transformation.

$$\frac{d^2\sigma}{dq^2 dv} \rightarrow \frac{d^2\sigma}{dx\,dy},$$

with

$$x = \frac{q^2}{2M_N v} \quad \text{and} \quad y = \frac{v}{E_i}.$$

$$\tag{29.6}$$

This is simple to do, after which the neutrino–proton cross-section can be written in terms of the neutrino–quark cross-sections (1) and (4) of (29.5), the proportions of the two being determined by the distribution of down quarks and antiup quarks inside the proton. This gives:

$$\frac{d^2\sigma}{dx\,dy} = f_d(x)\,\frac{d^2\sigma}{dx\,dy}\,(\nu_\mu + d \to \mu^- + u)$$

$$+ f_{\bar{u}}(x)\,\frac{d^2\sigma}{dx\,dy}\,(\nu_\mu + \bar{u} \to \mu^- + \bar{d}). \quad (29.7)$$

The neutrino–quark scattering is of a very simple point-like kind, and so the cross-sections just provide the usual factors for point-like scattering. Writing the above expression in terms of these factors leaves us with the differential cross-section for deep inelastic scattering, but with the structure functions expressed directly in terms of the distributions of quarks within the target nucleon.

$$\frac{d^2\sigma}{dx\,dy} = \underset{\substack{\uparrow \\ \text{neutrino–quark} \\ \text{point-like} \\ \text{scattering}}}{\frac{2MEG_F^2}{\pi}}\; \underset{\substack{\uparrow \\ \text{structure functions} \\ \text{expressed as} \\ \text{quark distributions}}}{[xf_d(x) + x(1-y^2)\,f_u(x)]}. \quad (29.8)$$

These expressions may be derived for ν and $\bar{\nu}$ scattering off both protons and neutrons. The average of the two gives the scattering of ν and $\bar{\nu}$ off a general nucleon target N consisting of a mixture of protons and nucleons. The ratio of $\bar{\nu}$N scattering to νN scattering allows the cancellation of point-like scattering factors leaving only a ratio of the quark distributions. By integrating over the variables x and y, the ratio of the total cross-sections is expressed in terms of the total fractional momentum carried by each species of quark.

$$\frac{\sigma^{\bar{\nu}N}}{\sigma^{\nu N}} = \frac{(P_u + P_d) + 3(P_{\bar{u}} + P_{\bar{d}})}{3(P_u + P_d) + (P_{\bar{u}} + P_{\bar{d}})}. \quad (29.9)$$

If quarks only are present inside the target nucleon, then this ratio will be $\frac{1}{3}$, and if antiquarks only are present, then the ratio will be 3. The measured value of $.37 \pm .02$ suggests a very slight presence of antiquarks from the sea.

The presence of the factors of 3 in the above ratio arises from the allowed helicity states of the νN collisions as described earlier, the integration over the y variable effectively being the integration over the allowed angles of νq scattering, thus favouring neutrino–quark scattering over neutrino–antiquark scattering (and vice versa for antineutrino scattering).

29.4 Electron and neutrino structure functions comparison

The quark content of the neutrino scattering structure functions may be obtained simply by comparing the differential cross-sections expressed in terms of quark distribution functions (29.7) with the same quantity expressed in terms of the structure functions. This comparison provides equalities such as:

$$\left.\begin{aligned} F_2^{\nu P} &= 2x\,(f_d(x) + f_{\bar{u}}(x)), \\ F_3^{\nu P} &= -2\,(f_d(x) - f_{\bar{u}}(x)), \\ F_2^{\nu N} &= 2x\,(f_u(x) + f_{\bar{d}}(x)), \\ F_3^{\nu N} &= -2\,(f_u(x) - f_{\bar{d}}(x)). \end{aligned}\right\} \quad (29.10)$$

By comparing the quark content content of the neutrino structure functions above with the quark content of the electron structure functions in (29.1), it is possible to predict a numerical relation between the neutrino and electron–nucleon scattering structure functions. The resulting relationship is:

$$F_2^{\nu N}(x) = \frac{18}{5}\,F_2^{eN}(x). \quad (29.11)$$

Experimentally, the relationship is found to hold very well, as illustrated in Figure 29.1. The significance of this is that the quark content of the target nucleon has been verified independently by two separate interactions and is thus that much more credible.

29.5 Sum rules

We can learn more about the roles played by the various quarks inside the proton by using expressions (29.1) and (29.10) to relate particular quark distribution functions to combinations of the observed structure functions. When integrated over the fractional momentum variable of the quark distributions, these relationships can reveal the total fractional momentum carried by each species of quark:

$$\left.\begin{aligned} \int x(f_s(x) + f_{\bar{s}}(x))\,dx &= \\ \int(9F_2^{eN}(x) - \tfrac{5}{2}\,F_2^{\nu N}(x))\,dx &= .05 \pm .18, \\[4pt] \int x(f_u(x) + f_d(x))\,dx &= \\ \tfrac{1}{2}\int (F_2^{\nu N}(x) - xF_3^{\nu N}(x))\,dx &= .49 \pm .06, \\[4pt] \int x(f_{\bar{u}}(x) + f_{\bar{d}}(x))\,dx &= \\ \tfrac{1}{2}\int (F_2^{\nu N}(x) + xF_3^{\nu N}(x))\,dx &= .02 \pm .03. \end{aligned}\right\} \quad (29.12)$$

The numbers show that, as expected, the strange quarks and the various sorts of antiquarks carry only a few per cent of the proton's total momentum. So the total contribution of 'sea' quarks is small. As

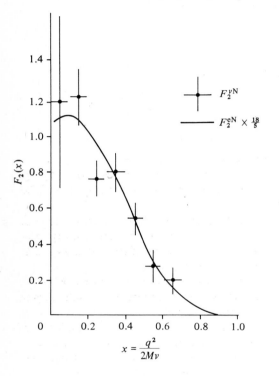

Fig. 29.1. The constant relationship between electron and neutrino scattering structure functions is verified by superimposing the two.

$$x = \frac{q^2}{2M\nu}$$

discovered in electron–nucleon scattering, the up and down quarks together carry about half of the protons' total momentum. Again the missing half of the protons' momentum is ascribed to the neutral gluons which are not affected by the weak interactions.

When we integrate the quark distribution functions (or, correspondingly, the structure functions) over the fractional momentum variable x, it is possible to derive 'sum rules' relating these quantities to physically significant numbers. For instance the Gross–Llewellyn-Smith sum rule measures the difference between the numbers of quarks and antiquarks in the target. As expected from the quark model, this number is measured to be approximately 3, as illustrated in Figure 29.2.

$$\tfrac{1}{2}\int_0^1 (F_3^{\bar{\nu}}(x)+F_3^{\nu}(x))\,\mathrm{d}x = N_q - N_{\bar{q}}$$

We have spent some time examining the structure functions of deep inelastic scattering as it is these which have provided us with a direct look inside the proton at its constituent quarks. All the observations are compatible with the standard model of spin-$\tfrac{1}{2}$ quarks with fractional charges. The big surprise is that these quarks carry only half of the proton's total momentum – the remainder presumably being carried by the gluons. The next step is to proceed to see if we can learn something of the dynamics of the interactions between quarks and gluons.

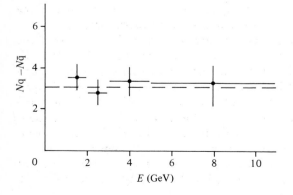

Fig. 29.2. The Gross–Llewellyn-Smith sum rule measures the difference between the numbers of quarks and antiquarks in the nucleon.

Part 7
Quantum chromodynamics – the theory of quarks

30

Coloured quarks

30.1 Introduction

The results of deep inelastic scattering experiments are able to tell us a lot about the nature of the quarks.

The scaling behaviour of the cross-sections indicates scattering off point-like quarks with relatively weak interactions between them at short distances.

The ratio of structure functions F_1^{eN}/F_2^{eN} supports the assignment of half-integer spin for the quarks.

The comparison of structure functions in electron and neutrino scattering reactions supports the assignment of fractional charges to the quarks.

The momentum sum rules in both electron– and neutrino–proton scattering suggest that quarks carry only about half of the total proton momentum. The other half is thought to be carried by neutral gluons, the quanta of the interquark force field.

This wealth of information on the structure of the proton was discovered between 1968 and the mid-1970s and represents an experimental triumph similar to the 1911 scattering experiments of Geiger, Marsden and Rutherford, which established the nuclear picture of the atom. In both cases, experimental observation led the way to the development of theories describing the phenomena.

Just as in 1913 Bohr's early quantum theory of the atom described Rutherford's discoveries, the formulation in 1973 of quantum chromodynamics (QCD) was put forward as a description of the behaviour of the quarks inside the proton. Pressing the analogy further, just as Bohr's description of the atom was an extension of the quantum theory propounded earlier by Planck, so QCD is an application of the ideas of gauge field theory developed in the 1960s.

QCD was proposed in 1973 by Fritzsch, Leutwyler and Gell-Mann (the last of whom, appropriately enough, was one of the original proponents of the quarks in 1963), although a similar idea had been put forward in 1966 by Nambu. The basic idea is to use a new charge called colour as the source of the interquark forces, just as electric charge is the source of electromagnetic forces between charged particles.

30.2 Colour

Soon after the proposal of the quarks, it was realised that the suggested quark content of some particles clashed with one of the most fundamental principles of quantum mechanics. The Pauli exclusion principle states that no two fermions (particles with half-integer spin) within a particular quantum system can have exactly the same quantum numbers. However, the proposed contents of some particles consist of no less than three identical quarks. For instance, the doubly charged, spin-$\frac{3}{2}$ resonance Δ^{++} must consist of three 'up' quarks, all with their spins pointing in the same direction, Figure 30.1(a). Similarly, the famous Ω^- particle, the discovery of which first confirmed the validity of the $SU(3)$ flavour scheme, must consist of three strange quarks, Figure 30.1(b). These two examples seem to contradict the rules of quantum mechanics.

One immediate way out of this problem is to suggest that the quarks are not fermions at all but, instead, are spinless or integer-spin bosons. However, it was realised early on that only fermionic quarks can account for the spins of the observed hadrons and subsequent observations, say, of structure functions in deep inelastic scattering, have always supported this conjecture. In fact, the mathematical statement of the Pauli exclusion principle deals in terms of the symmetry of the wavefunction which is the total description of the quantum-mechanical system. The statement that no two fermions can have exactly the same quantum

numbers in a particular system is equivalent to the statement that the wavefunction describing a system of fermions must be antisymmetric (i.e. it must change sign) on the interchange of any two of the constituent fermions. The wavefunction which describes a hadron made up of three quarks consists of at least three factors; one describing the positions of the quarks; one describing the spins of the quarks; and one describing the flavours of the quarks. The product of these three factors gives the overall wavefunction:

$$\psi_{TOTAL} = \psi_{SPACE} \times \psi_{SPIN} \times \psi_{FLAVOUR}.$$

For particles such as the Δ^{++}, all quarks have the same flavour, and so the flavour factor of the wavefunction is obviously symmetric under the interchange of any two quarks. The same is true of

Fig. 30.1. The quark content of the Δ^{++} in (a) and the Ω^- in (b) consists of three identical quarks, which would seem to contradict Pauli's exclusion principle. The introduction of colour distinguishes the quarks and preserves the principle.

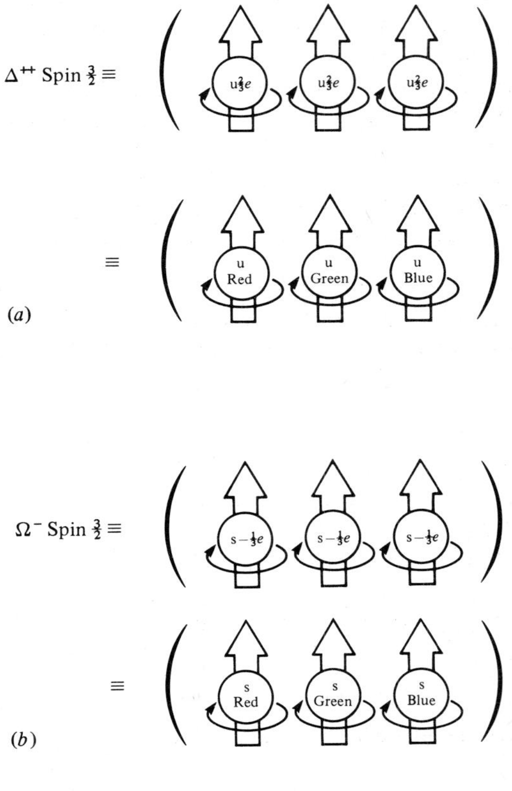

(a)

(b)

the spin factor because all quark spins are the same. Because the spins of the quarks add up to give the total overall spin of the particle, it means that there is no orbital angular momentum belonging to the three quarks. This implies that the quarks are positioned symmetrically so that the space factor is symmetric under the interchange of any two quarks. As all the individual factors are symmetric, the total wavefunction must be symmetric and the combination of the three quarks seems to violate the Pauli exclusion principle.

In 1964, Greenberg and, later, Han and Nambu, suggested that the quarks would have to carry another quantum number which would distinguish otherwise identical quarks and so avoid the demands of the Pauli exclusion principle. This new quantum number they called colour, although it should be stressed that this new property has nothing to do with the normal meaning of the word colour, it is just a label. The total wavefunction will now be multiplied by a new 'colour factor'.

$$\psi_{\text{TOTAL}} = \psi_{\text{SPACE}} \times \psi_{\text{SPIN}} \times \psi_{\text{FLAVOUR}} \times \psi_{\text{COLOUR}}.$$

The colour hypothesis is that each of the otherwise identical quarks has a different colour assigned to it and this makes the colour factor, and so the total wavefunction, antisymmetric under interchange of two quarks. The quark model is thus reconciled with the Pauli exclusion principle at the expense of introducing a new quantum number to differentiate between the quarks. Because there are three quarks inside the proton, three quark 'colours' are needed

Table 30.1. *A table of the flavours and colours of quarks*

Flavour \ Colour	Red	Green	Blue
$u\left(\dfrac{2}{3}e\right)$	u_r	u_g	u_b
$d\left(-\dfrac{1}{3}e\right)$	d_r	d_g	d_b
$s\left(-\dfrac{1}{3}e\right)$	s_r	s_g	s_b

to distinguish them uniquely. Say, red, green and blue. Each of these label the three quarks inside the Δ^{++} and the Ω^- as shown in Figure 30.1(*a*) and (*b*). So the net effect of the introduction of colour is to triple the number of quarks; each of the flavour must come in three colours. This is illustrated in the quark table, Table 30.1.

Obviously, tripling the number of the supposedly fundamental quarks is rather against the spirit of the model, which attempts to make do with as few basic components as possible. So the above arguments for colour, although theoretically compelling, had to be demonstrated by more-direct means before the colour quantum number became established as the physical reality on which a theory of quarks could be based. Happily, these more-direct means are readily evident.

One piece of evidence supporting the hypothesis that quarks come in three colours is provided by the decay of the neutral π^0 meson into two photons, see Figure 7.2. In the quark model, this decay rate is calculated by adding up all the possible varieties of quarks which can act as intermediate states in the decay. The experimental decay rate can be matched by the theory to within a few per cent if the quarks are assumed to come in three different colours. If only one colour of quark is allowed then the answer turns out to be a factor of 9, too small (the number of quarks entering the decay rate formula as a square).

A second piece of evidence for the existence of three different colours of quark is provided by the electron–positron annihilations. In these events, the electron and positron annihilate each other to produce a virtual photon loaded with energy. This virtual photon may then decay into either a muon and an antimuon or into a shower of hadrons. The shower of hadrons is taken to be the end result of the initial production of a quark–antiquark pair which subsequently transforms into the conventional hadrons, see Figure 30.2.

The ratio of the cross-sections for these events is a very significant quantity which will be discussed further in Part 8. Suffice to say at this stage that the ratio is proportional to the sum of the squares of the quark charges, and that only if each flavour of quark comes in three different colours does the number predicted by the quark model agree with the number observed experimentally.

So three separate pieces of evidence attest to the physical reality of the colour quantum number: quark 'spectroscopy' (i.e. how quarks build up the known hadrons); the π^0 decay; and e^+e^- annihilations. Two immediate questions arise from this reality. Firstly, if quarks carry these colour charges, is it possible to discover observable hadrons which also carry them? Secondly, given the existence of colour, what is its purpose? Can it form the basis of a theory of quark interactions?

30.3 Invisible colour

The observability of colour and the quark structure of matter are intimately linked. In fact, we will see that the introduction of colour corresponds to a formal method of categorising allowed quark structures. We saw in Part 2 how the observed hadrons fit into the multiplets generated by treating the up, down and strange quarks as the elements of the fundamental representation of the symmetry group $SU(3)$. However, we saw also that only very specific combinations of the fundamental representation (the quark flavour triplet) could generate the correct multiplets for the observed hadrons, see Table 30.2.

The introduction of colour provides a way of categorising which combinations of quarks and antiquarks are allowed to exist. The first step to realise is that the three colours of any one flavour of quark can be taken as the elements of the fundamental representation of a new symmetry group $SU(3)_C$. The methods of group theory then allow us to combine these fundamental representations (the quark colour triplet) into colour multiplets representing the different ways of combining all the colours of the quarks, see Table 30.3. The mathematics of $SU(3)_C$ is exactly the same as that for $SU(3)_F$, although it must be appreciated that these $SU(3)_C$ multiplets represent the various combinations of colour for a given flavour of quark and are completely distinct from the $SU(3)_F$ flavour multiplets. Just as it was necessary earlier with flavour multiplets to establish which of them could be taken to represent the observed hadrons, it is now necessary to repeat the exercise for the colour multiplets to see which of these are permitted.

The original purpose of the introduction of colour was to ensure that the combinations of quarks representing hadrons are multiplied by

Fig. 30.2. An electron–positron pair annihilates into a virtual photon (which may then disintegrate into a muon–antimuon pair), as in (a), or into a quark–antiquark pair (which then transforms into hadrons), as in (b).

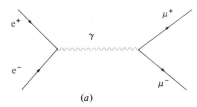

(a)

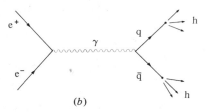

(b)

Table 30.2. *Certain combinations of the fundamental quark flavour triplets generate the observed multiplets*

Fundamental representation	Possible combinations	Multiplets generated	Seen
$q = \begin{pmatrix} u \\ d \\ s \end{pmatrix}$ 'flavour'	$q \times \bar{q}$ $q \times q \times q$ $q \times q$	$1+8$ $1+8+8+10$ 3^*+6	✓ ✓ ✗

Table 30.3. *Combinations of quark colour triplets generate multiplets*

Fundamental representation	Possible combinations	Multiplets generated	Seen
$q = \begin{pmatrix} r \\ g \\ b \end{pmatrix}$ 'colour'	$q \times \bar{q}$ $q \times q \times q$ $q \times q$	$1+8$ $1+8+8+10$ 3^*+6	? ? ?

factors which are antisymmetric under the interchange of two of the quarks (colours). Using mathematical analysis it is possible to show that some of the multiplets are antisymmetric, just as required, whilst others are symmetric, and so do not help us. The most obvious and simplest antisymmetric multiplet is the singlet. This observation is then elevated to a hypothesis for explaining the observed quark structure:

> All observed hadrons are colour singlets.

Under this hypothesis, $q \bar{q}$ and $q q q$ combinations are allowed because the series of colour multiplets generated includes a singlet. Also allowed are $q \bar{q} q \bar{q}$ and $q q q q \bar{q}$. These theoretically allowed combinations are referred to as exotic hadrons or baryonium states and they have long been the subject of experimental searches. Following several false alarms and tentative detections, their existence awaits experimental verification. Also under this hypothesis, combinations such as $q q$ or $q q q q$ are not allowed because those combinations of the fundamental representation of the colour group (quark colour triplet) do not generate a singlet combination. Experimentally, there has been no suggestion of their existence.

To summarise, let us recap the two parallel descriptions of quark structure, those of flavour and colour. Each combination of quarks generates a set of flavour multiplets and a set of colour multiplets. For certain of these combinations of quarks, the flavour multiplets will correspond to the observed hadrons and these combinations are identified as those which generate a colour singlet. All the observed hadrons are thought to be in colour singlet states. This scheme is summarised in Table 30.4.

The statement that all the observed hadrons are colour singlets is equivalent to saying that they are colourless. Just as flavour singlet states can carry no net electric charge or strangeness, the colour singlet states can carry no net colour. This can be understood simply by examining the colour combinations of the allowed quark structures:

$$q \bar{q} = \sqrt{\tfrac{1}{3}} \, (r\bar{r} + g\bar{g} + b\bar{b}),$$

$$q q q = \sqrt{\tfrac{1}{6}} \, (rgb - grb - rbg + gbr + brg - bgr).$$

The laws of quantum mechanics forbid us to say exactly what colour any one quark is at any one time, all we can say is that there is a certain probability of it being red or green or blue. However, what we can say is that in the singlet state of a $q \bar{q}$ combination the colour of the quark is exactly cancelled by the anticolour of the antiquark,

Table 30.4. *The colour and flavour multiplet structure of various quark–antiquark combinations and their postulated observability*

(a) $q \times \bar{q}$ quark–antiquark forms mesons like π^0, ρ, K, ψ, ...

COLOUR

$$\begin{pmatrix} r \\ g \\ b \end{pmatrix} \times \begin{pmatrix} \bar{r} \\ \bar{g} \\ \bar{b} \end{pmatrix}$$

$$\Downarrow$$

$$\mathbf{1} + \mathbf{8}$$

FLAVOUR

$$\begin{pmatrix} u \\ d \\ s \end{pmatrix} \times \begin{pmatrix} \bar{u} \\ \bar{d} \\ \bar{s} \end{pmatrix} \Rightarrow \begin{matrix} \mathbf{1} \\ + \\ \mathbf{8} \end{matrix}$$

$$\begin{pmatrix} \text{Yes} & \text{No} \\ & \\ \text{Yes} & \text{No} \\ & \text{Colour} \\ & \text{states} \end{pmatrix}$$

(b) $q \times q$ diquark states are not observed

COLOUR

$$\begin{pmatrix} r \\ g \\ b \end{pmatrix} \times \begin{pmatrix} r \\ g \\ b \end{pmatrix}$$

$$\Downarrow$$

$$\mathbf{3^*} + \mathbf{6}$$

FLAVOUR

$$\begin{pmatrix} u \\ d \\ s \end{pmatrix} \times \begin{pmatrix} u \\ d \\ s \end{pmatrix} \Rightarrow \begin{matrix} \mathbf{3^*} \\ + \\ \mathbf{6} \end{matrix}$$

$$\begin{pmatrix} \text{No} & \text{No} \\ & \\ \text{No} & \text{No} \\ & \text{Colour states} \end{pmatrix}$$

(c) $q \times q \times q$ three quarks form a baryon like the proton, neutron, Ω^-, Δ^{++}, ...

COLOUR

$$\begin{pmatrix} r \\ g \\ b \end{pmatrix} \times \begin{pmatrix} r \\ g \\ b \end{pmatrix} \times \begin{pmatrix} r \\ g \\ b \end{pmatrix}$$

$$\Downarrow$$

$$\mathbf{1} + \mathbf{8} + \mathbf{8} + \mathbf{10}$$

FLAVOUR

$$\begin{pmatrix} u \\ d \\ s \end{pmatrix} \times \begin{pmatrix} u \\ d \\ s \end{pmatrix} \times \begin{pmatrix} u \\ d \\ s \end{pmatrix} \Rightarrow \begin{matrix} \mathbf{1} \\ + \\ \mathbf{8} \\ + \\ \mathbf{8} \\ + \\ \mathbf{10} \end{matrix}$$

$$\begin{pmatrix} \text{Yes} & \text{No} & \text{No} & \text{No} \\ & & & \\ \text{Yes} & \text{No} & \text{No} & \text{No} \\ & & & \\ \text{Yes} & \text{No} & \text{No} & \text{No} \\ & & & \\ \text{Yes} & \text{No} & \text{No} & \text{No} \\ & \leftarrow \text{Colour} \rightarrow & & \\ & \text{states} & & \end{pmatrix}$$

and that in the singlet state of a q q q combination all the colours mix in equally to provide a 'white' baryon, i.e. one with no net colour quantum number. Under this scheme, the colour of the quarks is permanently hidden from us because all the allowed quark structures are colourless, colour singlets. The confinement of the quarks within the hadrons can accordingly be restated as the confinement of the colour quantum number.

At this point it is perhaps worth making clear that the confinement of quarks and of colour is really only a hypothesis which is occasionally reexamined, and quite properly so. In 1976, when some anomalous events were detected in electron–positron annihilations, Pati & Salam suggested that they may be the signals of unconfined quarks emerging freely and decaying into leptons rather than undergoing their forced transformations into hadrons. In fact, it was later realised that these anomalous events signalled the production of a new heavier brother of the muon. But, at least for a time, the free quark hypothesis was tenable. Similarly, in 1974, Matthews suggested that the discovery of the psi meson in electron–positron collisions could be interpreted as the creation of a colour meson (i.e. a particle from the 'No' column in Table 30.4 (*a*)), rather than the creation of the new 'charm' flavour.

In the end the evidence disproved this suggestion but it is by no means conclusively proved that such a coloured state will never be seen. Worthy of mention in passing is that the effects of new flavours are easily incorporated into the above picture and do not affect the colour of the quarks at all. The only effect of a new flavour is to generate bigger flavour multiplets to accommodate the greater number of quark combinations possible when an extra degree of freedom is present. The number of quark colours is always three, because that is the number required to distinguish the three valence quarks in each baryon.

31

Colour gauge theory

31.1 Introduction

The fundamental idea of QCD is that the colour 'charges' of the quarks act as the sources of the strong, so-called chromodynamic force between quarks, just as electric charge acts as the source of the electromagnetic force between electrically charged particles. As the quarks carry both colour and electric charge, they experience both the strong and electromagnetic forces, as well as the more feeble, weak and gravitational interactions. However, the chromodynamic force is by far the strongest in most regions of interest and so we are justified in examining it in isolation from the others.

In the terms of classical physics, the colour charges may be thought of as giving rise to a chromostatic force, just as electrical charges give rise to the electrostatic force given by Coulomb's inverse square law. Although there are great similarities between the two cases, the colour force will be a good deal more complicated. In chromodynamics there are three different colour charges between which the force must always be attractive to bind together the three different colours inside each baryon, and the force between colour and anticolour must also be attractive to bind together the quark and antiquark in each meson. However, despite these complications the analogy with the theory of electrodynamics is worth pursuing as far as possible.

Any theory of quarks, like any other fundamental theory, must be compatible with the laws of quantum mechanics and relativity and the most common approach to achieve this is relativistic quantum field theory. Both QED and the Glashow–Weinberg–Salam theory of the weak interactions are examples of a particular class of quantum field theory, namely a gauge theory. As this class has enjoyed such striking success in these other two areas, it must have seemed a reasonable candidate for describing the new area of the chromodynamic forces as well.

The technique of gauge theory is to describe the origin of the forces between particles in terms of some symmetry of nature. As we saw in Part 5, in QED the form of the interaction between electrons is dictated by demanding that the description of electron behaviour be invariant under arbitrary redefinition of the phases of electron waves. The demand for this invariance is simply an extension of the law of conservation of electrical charge. The redefinition of the phases of electron waves can be formulated mathematically as applying the transformations of some symmetry group to the electron wavefunction. The resulting formulation of QED is summarised in Figure 31.1.

31.2   The formulation of QCD

It will be useful to bear in mind the step-by-step summary of the formulation of QED because the formulation of QCD is remarkably similar. The first step is to identify the symmetry thought to be the fundamental origin of the colour forces. This comes readily to mind.

We have already seen that the quark colour triplet may be used as the fundamental representation of the symmetry group $SU(3)_C$ and that the colour multiplet structure generated by this group gives an acceptable way of categorising the known hadrons (i.e. all are in colour singlet states). The fundamental symmetry of the colour force may then be taken as the invariance of nature under the redefinition of quark colours. This is eminently consistent with the philosophy of quantum theory because, just like the phase of the electron wave, colour is believed to be unobservable and so no real phenomena should depend on the convention which defines it.

The redefinition of quark colours is achieved by applying the $SU(3)_C$ group of transformations to

Fig. 31.1 A summary of the formulation of QED.

(1) A wavefunction describes the propogation of an electron

(2) The Lagrangian describes the wavefunctions of two electrons in interaction

$\mathscr{L}(\psi_1, \psi_2)$

(3) Gauge invariance demands that in any theory of electrons the Lagrangian must be invariant under the redefinition of the phase of the electron wave. This is represented by the action of some group of transformations on the Lagrangian.

$$\mathbf{G}\psi \rightarrow \psi', \quad \mathbf{G}\mathscr{L}(\psi_1, \psi_2) \rightarrow \mathscr{L}(\psi'_1, \psi'_2)$$

(4) Furthermore, it is possible to require that this be true independently at each point in space. This is called local gauge invariance.

$$\mathbf{G}(x)\psi \rightarrow \psi', \quad \mathbf{G}(x)\mathscr{L}(\psi_1, \psi_2) \nrightarrow \mathscr{L}(\psi'_1, \psi'_2)$$

(5) For the Lagrangian to remain invariant under this last operation, a new gauge field must be introduced.

$$\mathbf{G}(x)\,\mathscr{L}(\psi_1, \psi_2, A) \rightarrow \mathscr{L}(\psi'_1, \psi'_2, A')$$

(6) This gauge field communicates between the two electron fields their locally defined phase conventions.

$\psi_1 \rightsquigarrow \psi_2$

(7) Stated in more familiar language, the electromagnetic field mediates the force between two electrons. In quantum theory, this is described as the exchange of photons, the quanta of the electromagnetic field.

the quark colour triplet. Suppose we define the quark colours initially by the multiplet.

$$q \equiv \begin{pmatrix} r \\ b \\ g \end{pmatrix}$$

We may then choose to change our colour scheme by applying the $SU(3)C$ group transformations to this triplet. This will have the effect of mixing up the colours to provide three different combinations each with different proportions of r, g and b.

$$\mathbf{G}^{SU(3)}\, q = \begin{pmatrix} c_1(r\ b\ g) \\ c_2(r\ b\ g) \\ c_3(r\ b\ g) \end{pmatrix} \equiv \begin{pmatrix} v \\ y \\ o \end{pmatrix}$$

However, it is perfectly possible to label these new combinations as new colours, say violet, yellow and orange. The underlying physical requirement is that the theory describing the quark interactions does not depend on whatever 'colour coding' is chosen. In fact, the colour coding of the quarks may be different at each point in space and this will require the Lagrangian to be *locally* gauge invariant under the application of the $SU(3)_C$ group of transformations.

Just as in the other gauge theories, this requires the introduction of a new gauge field to communicate the local colour conventions from place to place. The required form of this gauge field is one representing massless, spin-1 gauge particles. These are the gluons which are thought to carry the chromodynamic forces between the quarks. Because the interaction of two quarks represents the combination of two colour triplets, the gluons must come in the colour multiplets representing all the possible combinations of the colours which are generated by group theory. More precisely, the gluons should represent the combination of a quark colour triplet and an antiquark anticolour triplet as the quantum numbers of the gluon must equal those of the quark lines flowing *into* it. The required quantum numbers can be obtained by reversing the flow of an outgoing quark line and using its equivalence with an incoming antiquark line. This is illustrated in Figure 31.2. In fact, the gluons form a colour octet with quantum numbers such as red–antigreeness and antiblue–redness.

In QED the photon is electrically neutral and so does not act as a source of electromagnetic fields.

This means that photons cannot interact amongst themselves directly. The only way they can interact is for each to dissociate into a virtual electron–positron pair or other charged particle pair which may then do the interaction for them. However, this has a much lower probability of occurrence than the direct interaction between electrons. In QCD, the

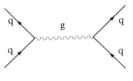

Fig. 31.2. Quarks interact by gluon exchange. The colour quantum numbers flowing into the gluon are equivalent to those of a q q̄ pair.

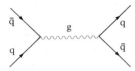

Fig. 31.3. (*a*) In QED, photons cannot interact directly, they must dissociate into an e^+e^- pair to do so at all. (*b*) In QCD, the coloured gluons can interact with each other directly.

QED

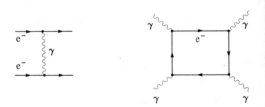

QCD

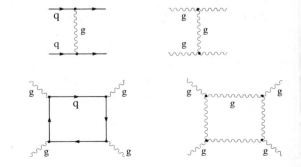

gluons carry colour and so give rise to their own colour fields. This means that they can interact amongst themselves directly. The two cases are illustrated graphically in Figure 31.3.

This difference between the two theories is a very fundamental one and has far-reaching consequences of great importance. The reason for the difference boils down to the number of charges in the theory. As we saw in Part 5, in the discussion on the Glashow–Weinberg–Salam theory of the weak force, QED is Abelian whilst the more complicated charges of the Glashow–Weinberg–Salam model

give rise to a non-Abelian theory. Similarly, in QCD the three charges give rise to a non-Abelian gauge group $SU(3)_C$, which means that the net result of a pair of gauge transformations depends on the order in which they are performed. Previously, we have demanded that the Lagrangian must be altogether invariant under gauge transformations, so the discrepancy arising from the order of transformations must be removed by adding some new feature to the Lagrangian. This new feature simply turns out to be a mathematical term describing the gauge fields in self-interaction. As a summary, Figure 31.4 shows a step-by-step approach to the formulation of QCD.

The picture of the hadrons painted by this theory is one of continual interchange of gluons between the constituent quarks which, as a result, continually keep changing colour but in such a fashion as to always maintain the hadron in its colour singlet state, see Figure 31.5.

One interesting consequence of the presence of gluon self-interaction in QCD is the possibility of the existence of particles made only of glue with no quarks. These are referred to as glueballs or gluonium states and are possible because when two colour octets are combined, a colour singlet state always results in addition to non-allowed colour multiplets, see Figure 31.6. Other combinations of

Fig. 31.4. A summary of the formulation of QCD.

(1) A wavefunction describes the propagation of a quark

$$\psi \quad \underline{\hspace{2cm}} \quad q$$

(2) The Lagrangian describes the wavefunctions of two quarks in interaction

$$\mathscr{L}(\psi_1, \psi_2) \equiv$$

$q^1 \qquad q_2$

(3) Gauge invariance demands that the Lagrangian must be invariant under the redefinition of the quarks' colour code.

$$\mathbf{G}^{SU(3)_c} \psi \to \psi', \quad \mathbf{G}^{SU(3)_c}, \ \mathscr{L}(\psi_1, \psi_2) \to \mathscr{L}(\psi'_1, \psi'_2)$$

(4) This gauge invariance may be required to hold locally

$$\mathbf{G}^{SU(3)_c}(x)\psi \to \psi', \quad \mathbf{G}^{SU(3)_c}(x) \ \mathscr{L}(\psi_1 \, \psi_2) \nrightarrow \mathscr{L}(\psi'_1 \, \psi'_2)$$

(5) The Lagrangian can remain invariant under this local group only if a new, self-interacting gauge field is introduced

$$\mathbf{G}^{SU(3)_c}(x)\mathscr{L}(\psi_1, \psi_2, \tilde{A}) \to \mathscr{L}(\psi'_1, \psi'_2, \tilde{A}')$$

(6) This gauge field communicates between the quarks their locally defined colour coding. In more-familiar terms, the quanta of the colour gauge field, the gluons, mediate the strong force between the quarks and also between themselves.

Fig. 31.5. The quarks inside the hadrons are bound together by continual gluon exchange.

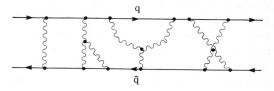

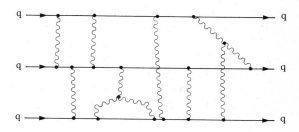

more than two gluons are also possible, giving rise to the possibility of a whole spectrum of gluonia. Their existence is currently the subject of lively debate, with some authors claiming as evidence of their presence certain irregularities observed in the quark model of the hadron spectrum, see Part 9.

Fig. 31.6. The non-Abelian nature of QCD allows the formation of particles made from gluons only.

$$8 \otimes 8 = 1 + 8 + 8 + 10 + 10 + 27$$

32

Asymptotic freedom

32.1 Introduction

What we have done so far, formulating a gauge theory of the colour force in analogy to the theory of the electromagnetic interaction, is all very well – but is it correct? Does it explain any of the features of the strong interaction which are observed in the real world? Only by this test, and not by its theoretical elegance or any other criterion, may it be accepted as correct. In particular, we are interested to see if the forces resulting from gluon exchange correspond to the behaviour of the strong interaction as observed in deep inelastic scattering. In these experiments, we saw that when the distances probed were very small, i.e. when the momentum transferred by the probe was very high, then the force between quarks is surprisingly weak and they behave rather like free particles. On the other hand, no free quark has ever been observed, so we may be sure that, over long distances, the force between quarks becomes increasingly strong.

Another and, as it turns out, related theoretical question is whether or not we can actually perform any meaningful calculations in QCD. In QED, calculations of quantities of physical interest are possible because the increasingly complicated higher-order processes become decreasingly important. This is due to the smallness of the electron–photon coupling constant, ($\alpha = \frac{1}{137}$). However, in QCD, the strength of the chromodynamic forces may require the quark–gluon and gluon–gluon couplings to be large (greater than one) and this would mean that the increasingly complicated higher-order processes become increasingly important. In this case, it is impossible to use the same mathematical techniques of perturbation theory to calculate quantities of physical interest. The whole point of perturbation theory is that the energy of interaction between the particles should be much less than the energy of the particles themselves. When the forces are strong this is not true and the method is not applicable.

Resolution of both the experimental and theoretical points mentioned above hinges on the far-from-obvious physical reality that the intrinsic strength of a force (the size of the coupling constant concerned) depends on the distance from which it is viewed. This dependence is in addition to the conventional spatial variation in the strength of forces as given, say, by the inverse square laws of classical physics.

A good example of this phenomenon is provided by the electromagnetic force. In classical physics the electrostatic force between two charged particles is given by Coulomb's inverse square law:

$$F = K \frac{N_1 e \cdot N_2 e}{r^2}.$$

In this formula, the intrinsic 'strength' of the force is fixed by the numerical value of the constant electric charge e. But when the distance separating the two particles becomes very small, then classical physics is no longer adequate and quantum-mechanical effects must be taken into account.

These quantum-mechanical effects can be described as the polarisation of the vacuum sea of virtual electron–positron pairs in the environment of an electric charge. In the region of an electron, say, the virtual positron is attracted towards it and the virtual electron is repelled away from it. This leads to a cloud of virtual positive charge shielding the 'bare' negative charge of the real electron, see Figure 32.1(*a*). The effect of this is that from a distance the effective negative charge is much reduced compared to its 'bare' value. The electric charge appearing in Coulomb's law is this shielded, effective charge.

The quantum-mechanical shielding effect is known as the renormalisation of the bare electrical charge and it can be calculated by evaluating the probabilities of occurrence of the various quantum-

mechanical processes such as the one illustrated in Figure 32.1(*b*). In fact, the numerical value of the probabilities of occurrence of the quantum-mechanical processes turns out to be infinity! In order to arrive at the finite value of the classical electrical charge, the 'bare' electric charge is taken to be negatively infinite so as to cancel out the quantum process and leave the finite value observed.

We can carry out a simple thought-experiment to investigate the quantum-mechanical behaviour of electric charge by considering the scattering of two electrons off each other at increasingly high energies. As the distance of approach decreases, the electrons begin to penetrate each other's virtual charge cloud and to experience each other's more negative bare charge. The change in the effective value of the electric charge is shown against distance of approach in Figure 32.2.

A parallel phenomenon exists in QCD. Just as the vacuum can be considered to be a sea of virtual electron–positron pairs, so too can it be considered as a sea of virtual quark–antiquark pairs and gluons, see Figure 32.3(*a*). The 'bare' colour charge on a single quark may then be shielded by the

Fig. 32.1. (*a*) Virtual electron–positron pairs shield the 'bare' electric charge at very short distances. This effect can be calculated by evaluating Feynman diagrams such as that shown in (*b*).

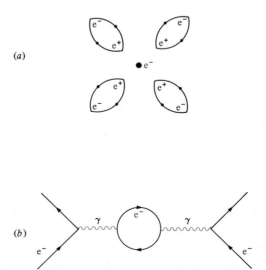

polarisation of this vacuum sea of virtual quarks, antiquarks and gluons. The resulting renormalisation of the bare colour charge may be calculated by evaluating the corresponding probabilities of occurrence of the various quantum-mechanical process, such as those shown in Figure 32.3(*b*). The essential new feature in the QCD case is the presence of gluon shielding, possible because of gluon self-interactions. Whereas in QED, the single variety of electron–positron shielding leads to a decrease in the effective electric charge compared to its bare charge, the presence of the gluon shielding effect in QCD provides a greater, opposite effect and leads to an increase in the effective colour charge relative to the original bare charge. Conversely stated, the effective strength of the colour charge on a quark appears to decrease as the distance from which it is viewed decreases, see Figure 32.2(*b*).

This phenomenon is qualitatively similar to the effect noticed in deep inelastic scattering: when the quarks are close together, the chromodynamic forces between them are weak; as the distance between them increases, so too do the forces. The term coined to denote this behaviour is 'asymptotic freedom', to denote the fact that when the inter-quark distances probed become asymptotically small (or, as in the original formulation, when the momentum of the deep inelastic probe becomes asymptotically high), then the chromodynamic forces disappear and the quarks become, effectively, free particles.

This remarkable property of QCD was discovered in 1973 by H. David Politzer of Harvard and independently by Gross and Wilczek of Princeton. Immediately, the feasibility of developing a field theory for the 'strong' interaction received a tremendous boost. For not only does the picture of the forces presented resemble their behaviour as observed in experiment, but also the demonstration that the 'strong' interaction coupling constant can, under certain circumstances, be small allows for the application of the traditional methods of perturba-

Fig. 32.2. (*a*) The apparent strength of an electrical charge as a function of the distance from which it is viewed. (*b*) The apparent strength of the colour charge on a quark.

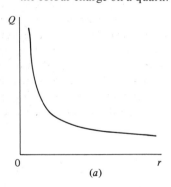

(*a*)

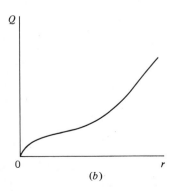

(*b*)

Fig. 32.3. (*a*) Virtual quark–antiquark pairs and gluons combine to 'antishield' the colour charges on the quarks. This effect is described by the Feynman diagrams shown in (*b*).

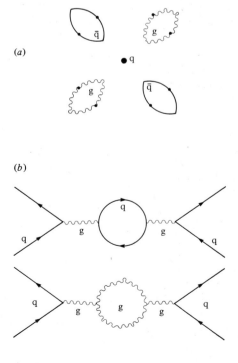

tion theory to calculate quantities of physical interest.

The position of QCD as the candidate theory of strong interactions was strengthened still further in 1974 when Gross and Wilczek went on to prove mathematically that only non-Abelian gauge field theories (of which QCD is an example) can give rise to asymptotically free behaviour. Also, they showed that this behaviour is possible only if there are a limited number of fermions in the theory (no more than 16 quark flavours in QCD) and if there are no Higgs particles involved in any form of spontaneous breaking of the $SU(3)_C$ symmetry. Thus if we decide that asymptotic freedom is a desirable feature of the theory of chromodynamic forces, certain other possibilities are decided for us automatically.

At this stage, it is desirable that we strengthen the credibility of QCD still further by the demonstration of the validity of the use of perturbation theory in calculating the effects of processes involving quarks and gluons. Fortunately, an example is close at hand.

32.2 Violations of scaling

The description of deep inelastic scattering in Part 6 represents a first attempt to gain some understanding of the interior of the proton. In this capacity it served us very well. For not only were we able to interpret the deep inelastic experiments as

the first dynamical evidence for the existence of the point-like quarks within the proton, but the implications of the experiments for the interquark forces provided us with the basic material for the formulation of QCD. Armed with this new theory we may now reexamine deep inelastic scattering and provide a more detailed explanation of the structure functions describing the constituents of the nucleon.

If the quarks were truly free particles then each would carry a third of the proton momentum (if we assume there are three valence quarks inside the proton). This would give the very simple form of proton structure function shown in Figure 32.4(a). However, we know that this is not the whole truth as the quarks are confined to within the dimensions of the proton, thus the uncertainty in their position can be no greater than $2r_p$. By Heisenberg's uncertainty principle this means that the uncertainty in their momentum must be at least

$$\Delta p \gtrsim \frac{\hbar}{2\,r_p} \approx 200 \text{ MeV}/c$$

and so the structure function tends to be smeared out as observed Figure 32.4(b).

A deeper understanding of deep inelastic scattering is possible when fuller data on the behaviour of the deep inelastic structure functions are examined. These data reveal that, far from being a constant shape for all values of momentum transferred (the original scaling hypothesis), the structure functions vary with it in a very well-defined fashion. This is shown in Figure 32.5(a) and (b). Interestingly, this diagram illustrates why scal-

Fig. 32.4. The expected behaviour of a nucleon structure function, for (a) free quarks, and for (b) confined quarks.

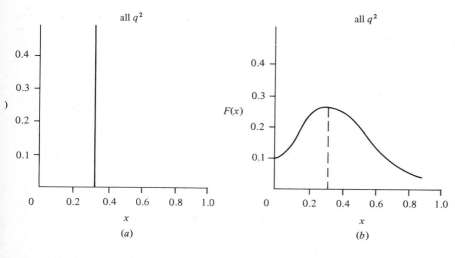

(a)

(b)

Fig. 32.5. (*a*) The violation of scaling behaviour; the nucleon structure functions vary systematically with momentum transfer squared. Values of *x* quoted are in fact the mid-points of ranges centred on those values. (*b*) The violation of scaling behaviour; the pattern of the variation of the nucleon structure functions.

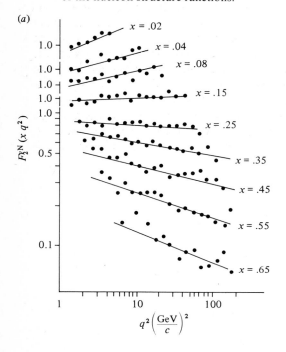

(*a*)

$F_2^N(x\,q^2)$

$q^2\left(\dfrac{\text{GeV}}{c}\right)^2$

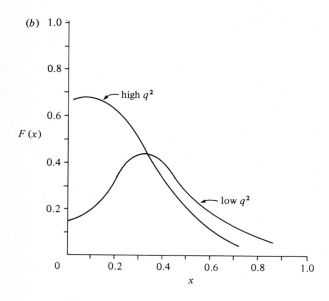

(*b*)

$F(x)$

ing was at first believed to be more exact than it really is. In the early experiments, the structure functions were examined for variations over only a limited range of q^2, predominantly in the mid-*x* regions – where there genuinely is no variation. The important variations in q^2 occur at low and high *x* values. Because of this, scaling was credited with more importance that its due. The fuller data show that it is not the constancy of the structure functions which is important, but their variation.

The variation of the structure functions is such that at low values of *x*, they increase with increasing momentum transfer, and that with high values of *x* there is a compensating decrease. This means that, as the momentum of the probe increases, it becomes more likely to hit a quark carrying a small fraction of the total proton momentum and less likely to hit a quark carrying a large fraction. This rather complicated behaviour can be understood by the application of the 'deep inelastic microscope technique' to the QCD picture of the proton.

As we have said, if there are no interquark forces then each valence quark will carry a third of the momentum of the proton. The corresponding structure function is shown in Figure 32.6(*a*). However, to confine the quarks inside the proton, we known that there must be some interquark forces – even if they do weaken in effect as the distances resolved by the probe decreases down from the proton diameter. In QCD, chromodynamic forces are mediated by the exchange of gluons between the quarks. This continual exchange of gluons transfers momentum between the quarks, so smearing out the deep inelastic structure function, Figure 32.6(*b*). As the momentum of the probe increases and the distance it resolves decreases, it begins to see the detailed quantum-mechanical sub-processes of QCD in the environment of the struck quark. For instance, what to a longer wavelength probe may have appeared to be a quark may be revealed to a shorter wavelength probe as a quark accompanied by a gluon, see Figure 32.6(*c*). What is more, the total momentum of the quark as measured by the long wavelength probe must now be divided between the quark and the gluon, leaving the quark with a lower fraction of the total proton momentum. So, as the momentum of the probe increases, the average fraction of the total proton momentum carried by the quarks appears to decrease – just as observed in Figure 32.5. As the momentum of the probe increases still further and its resolving distance becomes more minute, it may see the gluon radiated by the valence quark dissociating into a quark–antiquark pair from the vacuum sea. So there will appear to be even more quarks carrying very low fractions of the total proton momentum, Figure 32.6(*d*).

Using QCD, it is possible to calculate the probabilities of occurrence of these various quan-

tum-mechanical sub-processes and to derive the way the structure functions vary with the momentum of the probe. Unfortunately, this is a fairly complicated business and there is no direct way of comparing the predictions of QCD with the behaviour of the structure functions as described. Comparison is of course possible using rather more sophisticated descriptions of the structure functions. Suffice to say here that the behaviour of the structure functions is not inconsistent with the predictions of QCD but, on the other hand, it is not inconsistent with the predictions of other, non-asymptotically free field theories either. However, when viewed as only one piece of evidence amongst others, the QCD explanation of 'scaling violations' is generally considered to support the theory.

Fig. 32.6. The shorter the wavelength of the probe used, the more constituents are seen, each with a smaller fraction of the nucleons' total momentum.

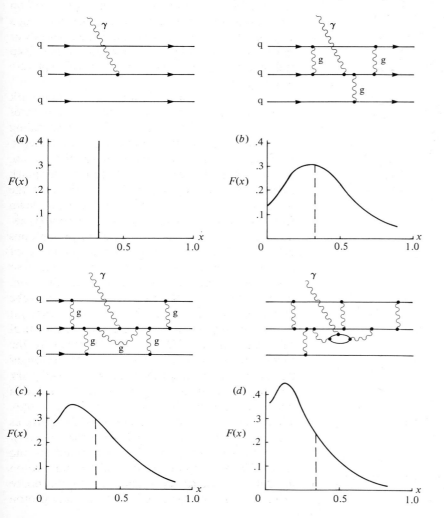

33

Quark confinement

33.1 Introduction

The fact that a single quark has never been observed has for years been the single greatest puzzle of elementary particle physics. No matter how energetically protons are collided together in the enormous accelerators at CERN and elsewhere, no quarks are seen to emerge in the debris. Many other varieties of particles are produced, but never any fractionally charged particles which may be identified with the quarks. This means that the forces which bind the quarks together are much stronger than the forces of the collision – which means that they are enormously strong. As an indication, we may note that the energies which bind the electrons into their atomic orbits are of the order of a few electron volts. The energies binding the protons and neutrons in the nucleus are of the order of a few million electron volts. Pairs of protons have been collided at energies of a few tens of thousands of millions electron volts and still no quarks are observed, which means the chromo-dynamic force between them must be at least that strong.

Not surprisingly, other more bizarre quark hunts have met with no success. Attempts have been made to detect the existence of fractional electric charges in all manner of materials from oysters (because they filter a large amount of sea water) to moon dust, with no convincing record of success. Because of the very delicate nature of the experiments, which are basically modern variants of

Milikan's oil-drop experiment, fractional charges are sometimes reported. But none of these have yet gained general acceptance (see Part 9).

These basic experimental facts have led theorists to conjecture that the quarks may be permanently confined within the hadrons as a result of the fundamental nature of the chromodynamic force. Just as Abelian QED gives rise to the inverse square law of Coulomb's law, it may be that the non-Abelian nature of QCD gives rise to a confining force which does not decrease with increasing distance. In fact, the corollary of asymptotic freedom is that the effective strength of the chromo-dynamic force increases as the quarks are drawn apart, a phenomenon known as infra-red slavery. It is not yet known whether QCD gives rise to infra-red slavery or whether, after a period of rising, the chromodynamic force tends to be a constant strength or even decreases as the quarks are separated. If the force does eventually begin to drop off, then the quarks will eventually be separable and confinement only a temporary phenomenon, apparent because our accelerator energies, although seemingly high to us, are not high enough. The various possibilities are shown in Figure 33.1

The major hindrance to a straightforward examination of the confinement problem is the difficulty in developing a mathematical description of strong forces. The method of perturbation theory used in QED and in the asymptotically free regime of QCD is valid only because the forces are weak. Attempts have been made to develop other methods such as one which divides the space–time con-

Fig. 33.1. Possible behaviours of the chromodynamic force at large distances.

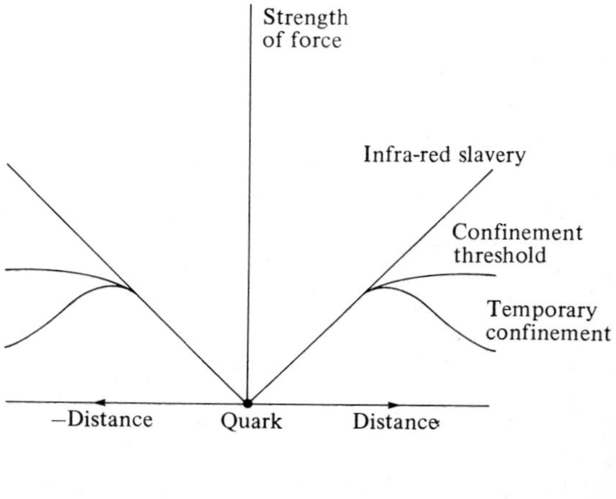

tinuum into a lattice of discrete points (so-called lattice gauge theory). But no totally satisfactory description has been achieved.

Instead, we will have to content ourselves with an intuitive picture of how the non-Abelian nature of QCD may give rise to the confinement mechanism. As usual, we start off with the familiar case of electrodynamics. The field lines joining two charges spread out to infinity in a spherical fashion. As they are drawn apart the field lines become more spread out. Because the density of field lines at any point is related to the strength of the electrostatic force at that point, this means that the force decreases as the separation increases, see Figure 33.2.

Consider now what may happen in QCD to the chromodynamic force between the quark and antiquark in a meson. The chromodynamic field lines would like to spread out like the electrodynamic ones, but because the non-Abelian nature of QCD gives rise to self-interactions of the gauge field, the field lines are drawn together instead. This is denoted in Figure 33.3 by field lines joining each other, as well as the quarks. As the quarks are separated, the field lines do not become more spread

out but, instead, are drawn out into a tube in which the density of chromodynamic force lines may be constant. This would lead to a constant force existing between the quarks like an infinitely elastic band. Eventually, as we put more and more work into increasing the separation of the quarks, the system will gain enough energy to promote a virtual quark–antiquark pair from the vacuum sea into physical reality. This will give rise to the creation of a new meson. So the energy we expend in attempting to separate the q q̄ pair has, in fact, resulted in the production of another meson, just as occurs in the high-energy collisions!

33.2 Quark forces – hadron forces

Having seen how QCD may provide an acceptable picture of the interquark forces, it is worth pausing to relate this picture to that of the forces between the observable hadrons, mentioned briefly in Part 2. These are the forces which bind the protons and neutrons together in the nuclei and, when the hadrons are in collision at high energies, produce the numerous secondary particles.

Fig. 33.3. Colour force lines between quarks are collimated into a tube-like shape and do not spread out as the quarks are separated. Eventually a single tube will split into two when the force applied has completed enough work.

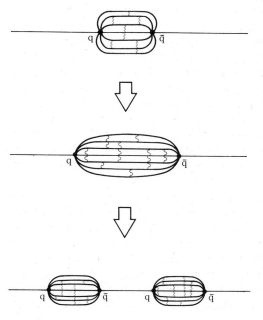

Fig. 33.2. Electric field lines spread out as the electric charges are separated.

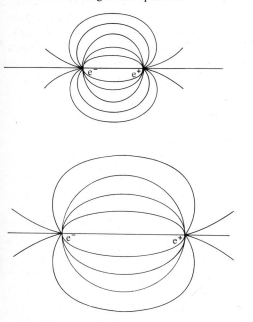

These forces are now seen as the van der Waals forces between hadrons. The van der Waals forces between atoms are the very feeble residual electrodynamic effects remaining after the electrons and nucleus have formed a net electrically neutral atom, Figure 33.4. Analogously, the van der Waals forces between hadrons are the chromodynamic effects remaining between colour singlet states once their colour constituents are bound together. Unlike the electrodynamic case, however, there is no guarantee that these 'secondary' forces will be weaker than the 'primary' interquark chromodynamic forces. This is because they are predominantly long-range phenomena, which is in the strong coupling regime of QCD.

Hadron collisions can be divided into either of two main classes. In the first are the diffractive collisions which are, in effect, glancing blows between the colliding particles. In Part 2 we saw that these form the vast majority of hadron collisions and that they can be described, albeit in a less than fundamental manner, by the Regge theory of collisions. In the QCD picture, these long-range collisions are complicated affairs involving multiple-gluon exchange with many sub-processes occurring, Figure 33.5. Because the forces are strong, there is no well-established method of describing the quark and gluon behaviour in these collisions. Indeed, there is no great motivation for

examining these collisions at the level of the details of quarks and gluons, as we are unlikely to be able to deduce much about the fundamental nature of the forces from such a complicated event. It is as if we were to attempt to study the electromagnetic force by observing collisions between complex atoms!

In the second main class of hadron collisions are the non-diffractive or 'head-on' events. Because they are much rarer than the diffractive events, they tended to be rather ignored prior to the development of QCD. Another reason for this perhaps was that they could not be described properly by Regge theory! However, at the levels of quarks and gluons, non-diffractive collisions are rather simple and so have become one of the centres of attention in the quest to understand more of the interquark forces.

Fig. 33.4. The analogy between the van der Waals force between atoms and the long-range colour force between the observed hadrons.

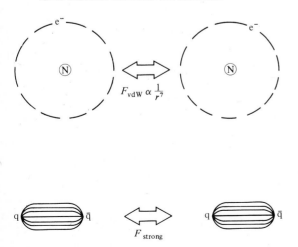

Fig. 33.5. Diffractive (glancing) hadron–hadron collisions are the result of complicated multi-quark and multi-gluon processes. In (a) the reggeon (R) exchange picture is shown. In (b) the QCD picture of the quark and gluon sub-processes is shown.

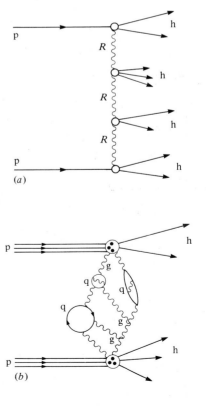

Because the hadrons collide head-one, it means that the quarks in each collision will approach each other very closely. The collisions are then thought to proceed predominantly by the exchange of a single gluon between two passing quarks, all the others acting as passive spectators, Figure 33.6. The result is that the interacting quarks are knocked violently sideways out of their parent hadrons. Of course, they do not emerge as free particles, the confinement mechanism dresses them up as hadrons, but the result is a jet of hadrons emerging along the directions of motion of the original quarks. These 'high transverse momentum' jets have recently been observed and study of their

behaviour is being used to determine the nature of the fundamental interactions behind them.

Jets occur also in other classes of high-energy collisions, such as electron–positron annihilations, where there are no complications due to spectator quarks. The events are altogether cleaner, as we will see in Part 8. But it is encouraging to note common phenomena in two very different circumstances, as this suggests a common, fundamental origin which we take to be the underlying dynamics of quarks and gluons.

Fig. 33.6. Head-on hadron–hadron collisions are described by simple quark and gluon processes, such as one-gluon exchange, which give rise to jets of hadrons emerging from the collisions.

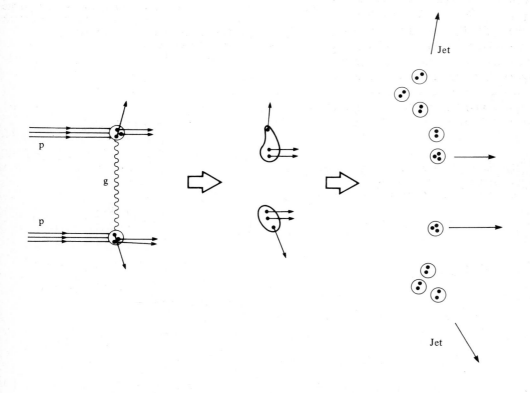

Part 8
Electron–positron collisions

Probing the vacuum

34.1 Introduction

Without doubt, the most fruitful class of experiments of the last decade has been that of electron–positron ($e^+ e^-$) collisions. For, just as the modern theories of the Glashow–Weinberg–Salam model and QCD have been formulated by generalisation of QED, the $e^+ e^-$ experiments allow study of both the weak and strong forces in comparison with the electromagnetic force – all are involved. So $e^+ e^-$ experiments provide a comprehensive laboratory for our ideas.

Another attractive feature of $e^+ e^-$ experiments is that because the electron and the positron are antiparticles, they often annihilate into a 'vacuum' state of pure energy. All the quantum numbers of the initial particles cancel and so we avoid the inhibiting effects of some conservation laws. The energy resulting from the annihilation (usually in the form of a virtual photon) is then free to sample the hidden, negative-energy content of the vacuum. It can be regarded quite literally as the photon of Figure 4.2 boosting a particle from its negative-energy sea to create, for instance, an $e^+ e^-$ pair or a $q\bar{q}$ pair. In this way $e^+ e^-$ experiments are ideal reactions in which to look for new particles.

An $e^+ e^-$ annihilation is also potentially rich in energy with which to create new particles. This is because, being antiparticles, the electron and the positron have exactly the same mass. If they are collided head-on, with equal and opposite momentum, the centre-of-mass of the reaction is stationary and all the energy is free to create particles. This is in contrast to accelerating electrons into a stationary target where momentum conservation requires that much of the energy be used in accelerating the target up to the required momentum, leaving only a fraction of the energy for particle creation. Another benefit of the stationary centre-of-mass of head-on collisions is that the angular distribution of the created particles can be measured directly and any significant asymmetries detected that much more easily.

Because the e^+ e^- pair can annihilate to a vacuum state, it means also that the reactions are very clean. There is no debris surviving from the initial state to mask interesting new effects or to confuse the experiments' detectors. This is in contrast to the deep inelastic experiments in which the photon–quark interaction has to take place in the presence of spectator quarks and leptons.

34.2 The experiments

Naturally, the many benefits of e^+ e^- collisions are balanced by some penalties. The basic requirement is to collide bunches of electrons with bunches of positrons head-on and this requires comparatively elaborate accelerator technology. The most serious drawback is the numbers of e^+ and e^- available in the bunches. The available flux limits the rate at which reactions can be observed, and this in turn limits the accuracy of the measurements obtained.

Another limitation is that the range of phenomena open to study depends upon the total energy of the collision. The total energy is just the sum of the e^+ energy and the e^- energy and, as the collision is head-on, all of it is available to create the mass of new particles. But whilst the e^+ e^- bunches are kept in orbit they emit their energy in the form of synchrotron radiation at a rate which rises rapidly with their energy (as the fourth power) and with the tightness of the bends (as the inverse of the radius of curvature). So, to achieve very high energies, it is necessary to build very large rings and to input a lot of radio frequency (rf) power to replenish the lost energy. So the search for new particles has lead to the construction of increasingly powerful accelerators. The progress of this construction in the past two decades is summarised in Table 34.1.

At the current frontier of e^+ e^- experiments

are the PETRA ring at DESY and PEP at SLAC, with the stupendous LEP at CERN promising an interesting end to the 1980s (see Chapter 37). But historically, it was the SPEAR ring at SLAC which provided such important advances during the mid-1970s and which may thus serve as the best illustration of e^+ e^- experiments, see Figure 34.1.

Electrons are created at the end of the two-mile-long linear accelerator tube. After some initial acceleration by electric fields, some of the electrons can be collided with a target to produce general particle debris from which the positrons can be filtered. Bunches of electrons and positrons are then accelerated down the tube by alternating electric fields and are then injected in opposite directions into the storage ring SPEAR. The bunches can be stored and accelerated in orbit in this ring for several hours by magnetic fields and rf power cavities, during which time e^+ and e^- bunches can be made to pass through each other at specified interaction regions. The individual e^+ e^- interactions are studied by many different types of detectors which are usually cylindrical or spherical distributions of sensors wrapped around the interaction region.

34.3 The basic reactions

It is useful to classify the various possibilities that can occur during an e^+ e^- collision. Firstly, there are the purely electromagnetic processes. 'Bhabba scattering' is the name given to elastic

Table 34.1. *Important e^+ e^- storage rings of the recent past and the near future*

Ring name	Location	Start	Centre-of-mass energy range (GeV)
SPEAR	Stanford, USA	1973	2.4–8.4
DORIS	DESY, Hamburg, W. Germany	1974	3.0–10.5 (12)
PETRA	DESY, Hamburg	1978	10–37 (45)
CESR	Cornell, USA	1979	8–16
PEP	Stanford, USA	1980	10–30
LEP	CERN, Geneva, Switzerland	1987–88	44–260

Fig. 34.1. An aerial view of the Stanford Linear Accelerator Center. The two-mile long linear accelerator is traversed by a highway. The historic SPEAR electron–positron annihilation ring is the small circular construction to the right of the largest hall. The much larger PEP ring, built more recently, is in the foreground and is indicated by the irregular track joining the ring of experimental halls. (Photo courtesy SLAC.)

$e^+ e^- \rightarrow e^+ e^-$ scattering. This can occur by either of the two Feynman diagrams of Figure 34.2(a) corresponding to the possibilities of photon exchange, and of annihilation into a virtual photon with subsequent reproduction of an $e^+ e^-$ pair.

The simplest electromagnetic process is muon-pair production $e^+ e^- \rightarrow \mu^+ \mu^-$, Figure 34.2(b), as this can occur only through the single-photon annihilation mechanism. The single photon must be virtual, as we remarked in Chapter 4, as it is impossible to conserve both energy and momentum. The energy of the initial two-electron state is always greater than $2 m_e$ but its momentum is zero; whereas the energy momentum relationship for real photons is $E = pc$.

If the energy of the $e^+ e^-$ pair is greater than $2m_\mu$ then the virtual photon can promote a muon pair from the negative-energy sea in the same way as it can reproduce an $e^+ e^-$ pair. Another possibility is the production of two real photons, Figure 34.2(c). This is allowed as the photons can emerge with equal and opposite momentum, so both energy and momentum can be conserved simultaneously.

Accurate measurement of these electromagnetic effects allows us to test the validity of QED at

Fig. 34.2. Possible electromagnetic effects following an e^+e^- collision. (a) Bhabba scattering. (b) muon-pair production. (c) two-photon production.

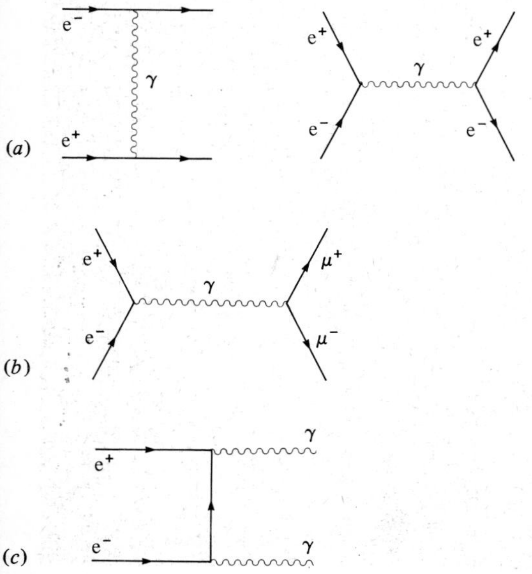

very high energies. This is done usually by measuring the angular distribution of particles emerging from the collision and comparing results with predictions. QED has always been found to be correct with the energies of reactions used so far. This can be reexpressed by saying that the leptons are effectively point-like down to 2×10^{-18} m (i.e. any sub-structure of the leptons is smaller than this).

The second class of $e^+ e^-$ collisions is comprised of those in which hadrons emerge in the final state and which indicate that the strong interaction is involved somewhere, see Figure 34.3(a). One of the most significant quantities in particle physics in the last decade has been the ratio R of the cross-section for $e^+ e^- \rightarrow$ hadrons to that for $e^+ e^- \rightarrow \mu^+ \mu^-$, measured as the energy of the collision varies. The significance of the ratio R is that it compares a reaction we understand very well (muon-pair production) with the class of reactions we wish to understand (hadron production) thus providing a very useful guide to our thinking about the unknown. Also, the ratio R is relatively straightforward to observe experimentally. Only two charged 'prongs' emerge in muon-pair production whereas, almost invariably, more emerge from a hadronic final state. So the ratio can be obtained by dividing the number of events detected with more than two prongs by the number with only two prongs, as measured during a given experiment.

Surprisingly, the ratio R is constant over large energy ranges, indicating that the complicated hadronic state is produced in much the same way as the simple muon pair. The virtual photon is probing the negative-energy sea of hadrons contained in the vacuum instead of that of electrons or muons. We will see how this can be given a clear interpretation in terms of quarks in the next chapter. Suffice at this stage to note that the hadrons cannot have been produced by the $e^+ e^-$ pair annihilating into a virtual gluon as the leptons have no colour and so have no connection with gluons whatsoever.

Before passing on, we must finally identify a third class of $e^+ e^-$ reaction involving the weak force. This results because the $e^+ e^-$ *do* carry leptonic isospin which allows them to annihilate into a virtual Z^0 boson. In fact, as we discussed during our look at the Glashow–Weinberg–Salam model, the photon γ and the Z^0 boson are simply the rather dissimilar quanta of the unified electroweak force. Thus we might expect weak interaction effects to come into the picture somewhere.

The virtual Z^0 boson is free to explore the negative-energy content of the vacuum just like the photon, see Figure 34.3(b). As a result we should expect to see some uniquely weak interaction effects (such as parity violation) creeping in at higher energies. This we will discuss further in Chapter 37. Until then, however, we will ignore these very slight effects. Most of our attention will focus on hadron production and the ratio R.

Fig. 34.3. Non-electromagnetic effects: (a) hadron production in the final state (the black blob); (b) annihilation into a Z^0 boson.

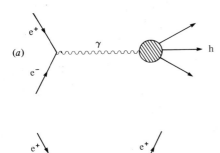

35

Quarks and charm

35.1 Introduction

The observation of scaling in deep inelastic scattering provides firm evidence for the interaction of the photon with point-like quarks inside the observed hadrons. So when we come to explain the process $e^+ e^- \rightarrow$ hadrons, the most likely picture is that of the virtual photon interacting with quarks rather than directly with complete hadrons, see Figure 35.1(a). The photon promotes a quark–antiquark (q q̄) pair from the vacuum, giving the quark and antiquark a kinetic energy depending on the initial collision energy. The q and q̄ must separate with equal and opposite momentum to maintain the net momentum of zero and in so doing are 'dressed up' into hadrons by the, as yet unknown, quark-confinement mechanism, see Figure 35.1(b). This may be viewed as the potential energy of the long-range attractive force between q and q̄ being used to promote extra q q̄ pairs from the vacuum.

35.2 The quark picture

Because the confinement stage of the process *always* occurs (at least assuming the permanently confined quark hypothesis), it enters the calculation of the process $e^+ e^- \rightarrow$ hadrons only as a final probability of one multiplying the underlying process $e^+ e^- \rightarrow$ q q̄. As the quarks are observed to be point-like and spin $\frac{1}{2}$, the process $e^+ e^- \rightarrow$ q q̄ is very similar to the process $e^+ e^- \rightarrow \mu^+ \mu^-$, the only difference being that the charges on the quarks are only some fraction of that on the muons. This explains the constancy of the ratio R mentioned earlier and displayed in Figure 35.2. The fundamental dynamics of the two processes are the same, so giving an R constant with energy, but their magni-

Fig. 35.1. $e^+e^- \rightarrow$ hadrons proceeds by a $q\bar{q}$ intermediate state, shown in (a). The transformation of this state into the observed hadrons involves the creation of more q q pairs (b).

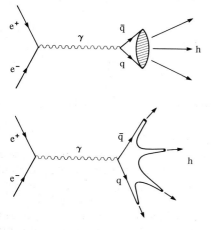

tudes differ by an amount equal to the ratio of the squares of the charges involved, see Figure 35.3. As several species of quark will be able to act as intermediaries to the creation of hadrons, and as the charge on the muon is 1, then R is equal to the sum of the squares of the quark charges.

Now the significance of R is gloriously apparent. It is a directly observable quark-counting opportunity which provides a measure of the number of quarks and their properties. For in-

Fig. 35.3. The value of the ratio R is equal to the sum of the squares of the quark charges.

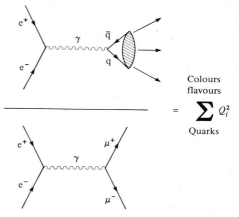

Fig. 35.2. The ratio R of the total hadronic cross-section to $\sigma(e^+e^- \rightarrow \mu^+\mu^-)$ as a function of the cm energy, E.

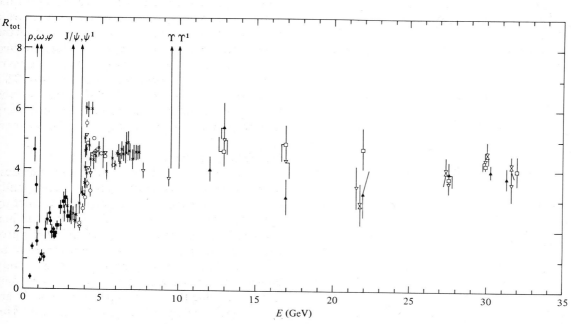

stance, in the simplest quark scheme with just three quark flavours; up ($\frac{2}{3}e$), down ($-\frac{1}{3}e$) and strange ($-\frac{1}{3}e$), the value of R is predicted to be

$$R_{uds} = \left(\frac{2}{3}\right)^2 + \left(-\frac{1}{3}\right)^2 + \left(-\frac{1}{3}\right)^2,$$

$$R_{uds} = \frac{2}{3}.$$

As we mentioned in Chapter 30 on QCD, the advent of the colour degree of freedom triples the number of quarks and so predicts the correct value at low energy, $R = 2$.

We have now explained the constancy of the ratio R but have not so far mentioned the very pronounced spikes which punctuate the picture. These shapes are highly reminiscent of the phenomena of resonance particles in Chapter 9 and this in fact is just what they are. At certain energies of the $e^+ e^-$ collision, the $q \bar{q}$ pair into which the photon transforms will have just the correct mass to stay intact as a single-meson resonance. This is signalled by a large increase in the probability of the event occurring compared with non-resonant 'background' $q \bar{q}$ production at neighbouring energies and this leads to the observed spikes in the cross-section. After its brief existence, the resonance particle will then decay by its usual mechanisms into the final-state hadrons observed.

The resonance particles produced are a select sub-set of the hundreds which have been observed in hadron–hadron reactions. The sub-set is defined by the quantum numbers of the virtual photon from which the resonances transform; spin 1, zero charge and strangeness. This defines the allowed quark content of the meson as being that of the quark–antiquark combinations in the vector nonet of $SU(3)$.

35.3 The advent of charm

In what was undoubtedly the most sensational experimental surprise of the 1970s, an extraordinary new resonance spike was discovered in $e^+ e^- \rightarrow$ hadrons at a collision energy of 3.096 GeV, followed quickly by the discovery of a similar spike at 3.687 GeV and a subsequently turbulent rise in the value of the ratio R to a new plateau, see Figure 35.2. The new resonance was denoted the ψ (psi) by Burton Richter and his colleagues at SLAC, who

observed the particle in $e^+ e^-$ annihilations, but it was also seen simultaneously as a resonance production phenomenon in the reaction $p + p \rightarrow e^+ e^- + X$ at Fermilab by Samuel Ting and his team who denoted it J. For their discovery both Richter and Ting were awarded the Nobel prize in 1976. Subsequently, ψ has become the accepted symbol for the particle at 3.097 GeV and ψ' for that at 3.687 GeV.

After a brief period of speculation, the correct interpretation of the ψ emerged. What had happened was that the increasing energy of the $e^+ e^-$ collision had become sufficiently large to create a new flavour $q \bar{q}$ pair. It had boosted a new heavier type of quark from its negative-energy sea in the vacuum. The ψ and ψ' were bound-state mesons consisting of the $q \bar{q}$ pair and, at energies above the threshold of its production, the pair could contribute to the ratio R, thus accounting for its observed rise.

This new flavour, called charm, had been anticipated in advance by the GIM theorists attempting to explain the behaviour of hadrons in the Glashow–Weinberg–Salam theory of the weak force. As we saw in Section 24.2, it was put forward as an explanation for the absence of strangeness-changing neutral currents. Also, it was able to complete an aesthetically pleasing matching between the numbers of fundamental leptons and fundamental hadrons. With the advent of charm, it became possible to group the leptons and the quarks into two generations, the second being simply a massive repetition of the quantum numbers of the first, see Table 35.1.

But the discovery of the charmed quark automatically implies the existence of a horde of new particles corresponding not only to the excited

Table 35.1 *The charmed quark completes two generations of the fundamental quarks and leptons*

Generation	First	Second	Charge
Quarks	u	c	$\frac{2}{3}$
	d	s	$-\frac{1}{3}$
Leptons	e^-	μ^-	-1
	ν_e	ν_μ	0

states formed from the $c\bar{c}$ pair, but also to all possible combinations of the charmed quark with the up, down and strange quarks in both mesonic and baryonic configurations and their excited states. In short, the $SU(3)$ flavour symmetry of the hadrons is enlarged to $SU(4)$. Mesons which combine the charmed quark with the up or down antiquarks are denoted the D mesons. These mesons carry explicit charm (i.e. have a non-zero charm quantum number), just as the K mesons carry strangeness. This is in contrast to the ψ itself which, being a $c\bar{c}$ combination, has the charm of its quark cancelled by the anticharm of its antiquark. There will also be mesons consisting of both charmed and strange quarks and antiquarks, denoted the F mesons, which are thus both charmed and strange. All the spin-0 mesons possible are generated by the hexadeciment of $SU(4)$ symmetry, which is illustrated in Figure 35.4. Similarly, there will be new patterns of baryons introducing charmed and strange baryons.

Fig. 35.4. The hexadecimet (16-plet) of spin-0 mesons generated by $SU(4)$ flavour symmetry. The familiar nonet of $SU(3)$ flavour is the middle $C=0$ plane.

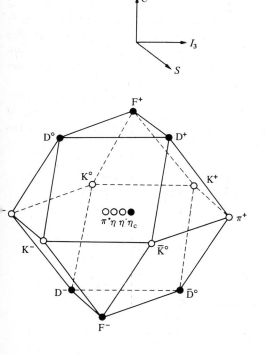

With the prospect of such a feast of new particles, both experimenters and theorists busied themselves in the 1970s in confirming the anticipated picture. It often became a race between experimental teams to be the first to detect a particularly tricky candidate, whilst theorists vied with each other to predict the masses and properties of the particles as accurately as possible. This led to a rapid advance in our understanding of quark behaviour and in the formulation of QCD.

35.4 Psichology

What remained a puzzle for some time was the extraordinary size of the resonance (its formation being some 3000 times more probable than the production of a non-resonant $q\bar{q}$ pair at neighbouring energies), and its extreme narrowness. The ψ has a width of only 0.002% of its mass compared with the ρs width of 20% of its mass. By Heisenberg's uncertainty principle this means that the ψ has a lifetime much longer than that generally associated with hadrons.

This unusual narrowness can be explained in terms of the inhibition of its preferred decay modes because of the masses of the charmed mesons. Its preferred decay mode would normally be expected to be into the charmed mesons $D^+ D^-$ or $D^0 \bar{D}^0$, proceeding by a quark line diagram rather like that for the decay of the ρ^0 into $\pi^+ \pi^-$ or $2\pi^0$, see Figure 35.5(a). However, the ρ^0 decay can proceed only because the ρ^0 mass of 0.77 GeV is larger than the mass of the two-pion state, 2×0.135 GeV. The ψ is so narrow because its mass is *less* than that of two charmed mesons. These mesons were detected well

Fig. 35.5. The obvious decay mode of the mesons (a) is not possible because the charmed mesons are too massive. The more complicated decay into non-charmed hadrons (b) takes longer.

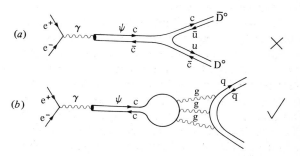

after the discovery of the ψ with a mass of 1.86 GeV, thereby requiring a particle with a mass of at least 3.72 GeV to produce them.

The ψ can decay only by rather sophisticated means. The $c\bar{c}$ pair has to annihilate itself into a state of three gluons which must then transform themselves into the observed hadrons by the mechanism of colour confinement similar to that practiced by the quarks mentioned earlier, see Figure 35.5(*b*). Intermediate states of one or two gluons are prohibited by the conservation laws of momentum and C parity respectively.

The ψ is but the most obvious of a whole family of mesons consisting of a $c\bar{c}$ pair. Some differ only in mass, the differences being due to the increased radial excitation energy of the $c\bar{c}$ pair. The ψ' at 3.687 GeV is the first of these and, like the ψ, is very narrow because it too lies below the D $\bar{\text{D}}$ threshold. The ψ'' is next at 3.77 GeV but this is a hadron of normal width (at 0.7% of mass) as it lies above the D $\bar{\text{D}}$ threshold (but only just!). Above this there are a number of other states.

However, some $c\bar{c}$ mesons have different spin, parity and C-parity assignments to that of the ψ, and these can be discovered only when the heavier members of the ψ family emit a photon, thereby allowing the $c\bar{c}$ pair to change its quantum numbers from those of the original virtual photon.

A typical process is shown in Figure 35.6 and a simplified version of the entire $c\bar{c}$ family is shown in Figure 35.7. To discover all these 'secondary' $c\bar{c}$ states, it is necessary to observe the energies of photons emerging from a process such as that in Figure 35.6. If one energy is preferred above all those possible, this is taken to indicate the mass difference between the heavy ψ-like particle (the energy of the e^+e^- collision) and the secondary $c\bar{c}$ state with different spin or parity assignments.

To achieve this prodigiously detailed particle-hunting task, experimenters at SLAC built a novel

photon detector nicknamed the crystal ball. This consists of a spherical array of sodium iodide crystals pointing towards its centre which is co-located with the interaction region. The sodium iodide crystals are monitored by photomultipliers which can measure the energy deposited in the crystal by an incident photon, see Figure 35.8.

Readers familiar with atomic physics will recognise the pattern of Figure 35.7 as being very similar to the energy level structure of the hydrogen atom. This similarity is understandable because the $c\bar{c}$ pair have bound themselves together into an exotic sort of elementary particle atom. Recognition of this phenomenon provided an enormous opportunity for particle physicists because such an atomic arrangement of the relatively heavy charmed quarks can be described by well-understood non-relativistic quantum mechanics. The force between the quarks can be formulated as a potential acting in the vicinity of a colour charge, just as in classical electrodynamics an electric potential surrounding an electric charge gives rise to Coulomb's force law between charges.

Fig. 35.6. If a heavy ψ-like state emits a photon, c $\bar{c}$ mesons with new quantum numbers are created.

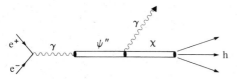

Fig. 35.7. The experimentally observed spectrum of $c\bar{c}$ mesons resulting from the different values possible for the spin and the orbital angular momentum of the constituent quarks. In this notation S refers to zero orbital angular momentum, *P* refers to the value $\hbar$, and *D* refers to the value $2\hbar$.

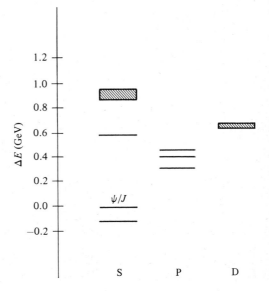

The particular form of the potential will determine the splitting of the energy levels or, in the $c\bar{c}$ case, the mass differences between mesons. As these can be measured experimentally with great accuracy, this can be used to provide a detailed picture of the force between the quarks.

The form of the potential arising from a colour charge which is found to give the most satisfactory match to the spectrum of mass levels is

Fig. 35.8. The crystal ball detector at SLAC. Numerous photomultiplier tubes bristle from the surface of the spherical container. They are monitoring the sodium iodide crystals mounted in the interior which detect the photons originating from the interaction point. (Photo courtesy SLAC.)

one which combines a simple Coulomb law at short ranges (one corresponding to single gluon exchange in the asymptotically free regime) with an attractive potential rising linearly with range at longer ranges, giving rise to the ever-increasing forces of quark confinement. The theoretical pattern of $c\bar{c}$ mass states generated by this potential is shown in Figure 35.9(b) and, in comparison, the energy levels of positronium, the bound states of e^+e^- arising from the Coulomb potential between the two electric charges, is shown in Figure 35.9(a).

So the masses of the $c\bar{c}$ mesons (sometimes referred to as the spectrum of charmonium) provide direct support for the QCD picture of interquark forces containing both asymptotic freedom at short ranges and confining forces at longer ranges.

Fig. 35.9. The spectrum of energy levels expected from the familiar electric potential is shown in (a). In (b), is shown the spectrum generated by the proposed form of the interquark potential It is a much closer match to the observed spectrum.

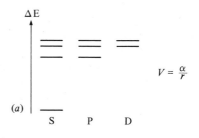

$$V = \frac{\alpha}{r}$$

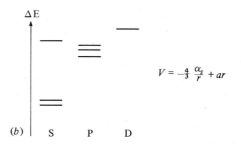

$$V = -\frac{4}{3}\frac{\alpha_s}{r} + ar$$

Fig. 35.10. A magnified bubble chamber photograph of charmed particles decaying. In the top half, a positively charged charmed meson (track entering from left) decays into three other charged particles. In the lower half, an invisible, neutral charmed meson decays into a pair of charged particles.

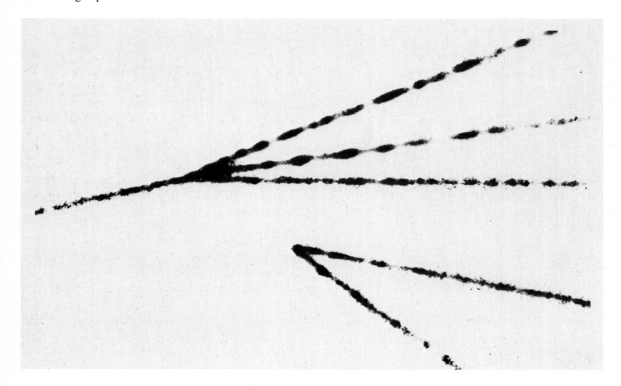

35.5 Charmed particles

For several years after the discovery of the ψ, experimenters sought the scores of particles which should be expected to carry explicit charm and their various excited states with ever-increasing spins. These were a lot harder to dig out of the experiments, as they could be found only by searching amongst the final-state hadrons for particular combinations at given masses. When a lot of debris is present in the final state and when the decays of the sought-for particle are uncertain, this is a tricky business. Eventually, a respectable roll-call of the particles was built up which supports their categorisation by $SU(4)$ flavour symmetry.

Like the strange particles, charmed particles will decay by the strong force, emitting pions until they arrive at the lowest-mass charmed state. Charm is conserved by the strong force and so this state, for example a D meson, is obliged to decay by the weak force. This it does by emitting a virtual W boson which changes the flavour of the emitting quark. This is an extension of the Cabbibo hypothesis of Chapter 18. The charmed quark will prefer to turn into a strange quark rather than an up or down, and this is signalled by the presence of a high proportion of strange particles amongst the decay products of the Ds. These strange particles must then decay either to non-strange mesons or directly into leptons by another weak interaction process. Thus the decay of the charmed particle is a complex laboratory of weak decays involving as many as three in succession, see Figure 35.10.

In summary, the discovery of charm has enabled us to find out a great deal about the strong force between quarks, as carried by the gluons of QCD, by studying the spectrum of $c\bar{c}$ mesons. The decays of the charmed D and F mesons has confirmed our understanding of the weak decays of hadrons as contained in the Glashow–Weinberg–Salam theory.

36

Another generation

36.1 Introduction

Soon after physicists had digested the consequences of the ψ mesons and the charm scheme, the discovery of yet another particle threatened them with elementary particle indigestion. In an experiment similar to Ting's discovery of the ψ, Leon Lederman and his team at Fermilab discovered a new particle in the reaction.

$$p + N \rightarrow \mu^+ \mu^- + X.$$

Lederman and his colleagues observed that this reaction was enhanced slightly for a $\mu^+ \mu^-$ pair mass of 9.46 GeV compared to its generally declining probability over the neighbouring range, see Figure 36.1. This was taken as the signal of a new, very massive meson resonance consisting of yet another flavour of quark bound to its antiquark. The new meson is denoted by upsilon Υ and its new constituent, the bottom quark, b (after a spirited but doomed effort on the part of a romantic school to get it called beauty).

36.2 The upsilon

This interpretation was by no means certain at the beginning and the pN experiment is by no means an ideal reaction in which to study the particle. This is because the hadronic debris X confuses the final state, and the fact that the very massive $\mu^+ \mu^-$ pairs are relatively rare makes it difficult to obtain accurate statistics. If its interpretation were correct, then it should be produced also in $e^+ e^-$ annihilations exactly like the ψ and so this was the obvious way to examine it in more detail. The trouble was that with its mass at 9.46 GeV, the Υ lay above the energy range of the SPEAR ring at SLAC and *below* the range provided by the new PETRA ring opened at DESY in 1978. Doubtless, the high-energy planners thought that no divine guiding hand would deal such a low card as to stick a particle between 8.4 and 10 GeV. However, it was vital that the Υ be investigated in the uncluttered environ-

Fig. 36.1. The $\mu^+\mu^-$ mass spectrum in pN collisions, containing the tell-tale bump of the upsilon.

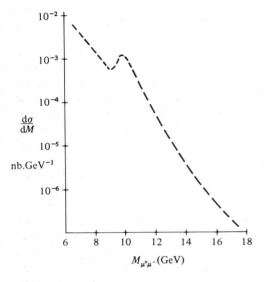

ment of e^+e^- annihilations and so the energy range of the DORIS ring (PETRA's predecessor at DESY) was tweaked to give just enough energy to reach the Υ.

The e^+e^- experiments confirmed that the Υ was indeed a $(b\bar{b})$ bound state and confirmed also the existence of its radially excited relatives, Υ' at 10 GeV and Υ'' at 10.40 GeV, see Figure 36.2. The width of the states was much harder to establish than that of the Υ as the energy resolution of the storage ring is not as accurate at the very end of its energy range as in the middle. The best value for the Υ width is about 0.005% of its mass, which indicates that it too, like the Υ, has its preferred decay mode (into explicit bottom mesons) suppressed. It too must annihilate the bottom of its quark with the anti-bottom of its antiquark into a state of three gluons which will then transform into non-bottom hadrons. The full significance of this will be seen later in Chapter 37. From measurement of the Υ width, it is possible to deduce that the most likely charge of the bottom meson is $-\frac{1}{3}$, which establishes it as a more massive successor to the down and strange quarks. The spacing of the masses of the Υ and Υ' can be calculated in the same way as the spectrum of ψ states. The experimental value observed supports the form of the interquark force as described by the potential of Figure 35.8(*b*).

The existence of yet another flavour of quark of course means that there must exist an entire new family of mesons with explicit bottom for all the various values of isospin, strangeness and charm discussed previously. The $SU(4)$ flavour symmetry is enlarged to $SU(5)$ so that the basic multiplet of

Fig. 36.2. Evidence for the Υ and Υ' from the total cross-section for $e^+e^- \rightarrow$ hadrons.

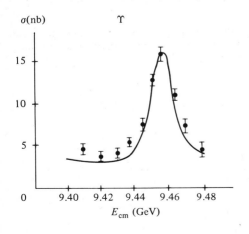

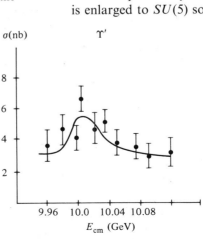

spin-0 mesons is now expanded from the hexadeci-ment of Figure 35.4 to a 25-plet. Similarly, baryons with non-zero bottom will augment all the baryonic multiplets. Detection of explicit bottom particles is even harder than that of naked charm as they are much more massive and thus require high-energy collisions. These will contain more debris in the final state from which the suspected decay products of the bottom mesons must be sorted. Despite these difficulties recent experiments have detected explicit bottom particles, as shown in Figure 36.3. Eventually, detailed experimental evidence on bottom particle spectroscopy should provide confirmation of the quark dynamics formulated in the context of the charm spectrum.

In some ways, just the existence of the Υ meson and bottom quark is of more significance than the details of its properties. For there is no place for the bottom quark in the first two generations. This suggests that it is the herald of a third generation containing yet another quark (denoted, naturally, the top quark) and a new lepton and its neutrino. Indeed, simultaneous with the discovery of the Υ, evidence for a new lepton was already mounting.

36.3 The tau heavy lepton

In 1975, at the time of the $e^+ e^-$ charm

Fig. 36.3. Experimental evidence for the production of explicit bottom hadrons. An excess of electron production at the beam energy of an Υ state suggests that it is decaying into bottom mesons which then produce the electrons in their own weak decays.

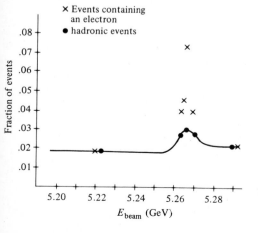

experiments, a team of physicists led by Martin Perl, also working on the SPEAR ring at SLAC, reported the existence of 'anomalous μe' events occurring in $e^+ e^-$ reactions. They suggested that these might signal the existence of a new heavy lepton, denoted τ. The 'anomalous μe events' are reactions of the form:

$$e^+ e^- \rightarrow e^\pm \mu^\mp + \text{nothing}$$

and the suggested origin of the final state is that of the separate electronic and muonic decays of the new intermediate pair of heavy leptons, see Figure 36.4.

It took some time to establish the truth of Perl's suggestion, due to several complicating factors. The most serious of these was that the energy threshold for the production of the $\tau^+ \tau^-$ pair is approximately 3.6 GeV (implying a mass for the τ of about 1.8 GeV). This of course, is very close to the threshold of 3.72 GeV required for the production of a charmed meson pair $D^0 \bar{D}^0$. As we know, these must decay by the weak interaction and so can quite easily be confused with tau heavy-lepton production and decay. However, in the case of charmed mesons, one would generally expect other hadronic tracks to be present. It is extremely unlikely that charmed-particle decays will give rise to the final state detected by Perl.

The problems were in ensuring that absolutely no other charged particles had been produced and had slipped past the detectors, or that the electrons and muons detected were in fact hadrons confusing the detectors (misidentification is always possible).

Eventually, Perl was able to place his identification beyond doubt, the final evidence for this

Fig. 36.4. Production of a $\tau^+ \tau^-$ heavy-lepton pair in $e^+ e^-$ annihilation gives rise to an 'anomalous' μe final state.

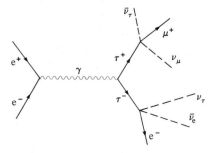

being the production of the μe states *below* the threshold for charmed meson-pair production (possible because of the slightly lesser mass of the τ). The evidence for this production is shown in Figure 36.5 which shows the growth of the process away from its theoretical threshold. The shape of the energy dependence also establishes the spin of the τ to be $\frac{1}{2}$, like that of the electron and the muon, and in contrast to the spin-0 D mesons. Since its confirmation, all the evidence has supported the identification of the τ as being a very massive copy of the muon (which is itself simply a massive copy of the electron). Like all leptons, the τ does not experience the strong force as do the quarks. However, the τ does have one new feature compared to the muon and the electron. Because of its large mass, it can

decay into hadrons and this means that it can add up to one unit to the ratio R, as defined previously, above its production threshold, over and above the value predicted by the charges of the quarks.

Apart from this new feature, the τ behaves exactly like its less-massive relatives during interactions. Despite its mass, it shows no deviation from point-like behaviour down to the current experimental limit of 2×10^{-18} m, and provides us with no hints that the leptons themselves may be composites of even smaller particles.

Further experiments have determined that, just like the electron and the muon, the tau has its own tau-type neutrino, and that tau number is conserved during weak interactions. The advent of the tau-neutrino completes the third generation of fundamental leptons, as shown in Table 36.1.

It is an amusing historical diversion to note that the first generation was begun in 1897 by J. J. Thompson with his discovery of the electron and concluded between 1963 (with Gell-Mann's introduction of the u and d quarks) and, let us say, 1968 (with their shadows being detected in deep inelastic experiments).

The second generation was begun in 1937 with Anderson's (mis)identification of the muon and completed in 1974 with the evidence for the charmed quark.

The third generation began in 1975 with Perl's first evidence for the tau and is still awaiting completion in the discovery of the sixth, top quark. As we shall see next, it will be surprising if this generation takes anything like the length of time needed to fill in the other two!

Fig. 36.5. The growth of $\tau^+\tau^-$ production from threshold as signalled by the ratio of candidate events (those containing an electron and some other charged particle only, eX) to known $\mu^+\mu^-$ production.

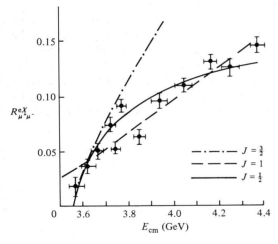

Table 36.1 *The generations of fundamental particles are almost complete. Only the top quark is missing*

Generation	First	Second	Third	Charge
Quarks	u	c	(t)	$\frac{2}{3}$
	d	s	b	$-\frac{1}{3}$
Leptons	e^-	μ^-	τ^-	-1
	ν_e	ν_μ	ν_τ	0

37

$e^+ e^-$ today and tomorrow

37.1 Introduction

Since 1978, attention has focussed on the PETRA $e^+ e^-$ ring at the DESY Laboratory, near Hamburg in Germany, which has looked at $e^+ e^-$ collisions up to the highest energy attained so far, 38 GeV. Figure 37.1 shows the basic layout of PETRA and how the previous generation storage rings are used as preaccelerators. Each of the five experiments around the ring consists of a very large array

Fig. 37.1. The layout of the PETRA ring at DESY.

of the various types of detectors wrapped around the interaction regions, each configuration being tailored specifically to search for the effects sought by that particular experiment. As many as nine separate institutes collaborate to run one experiment.

In one sense, the major result from this vast effort is rather a negative one. The value of the ratio R is much the same at 35 GeV as at 10 GeV, indicating that no new intermediate states are being produced by the virtual photon in this energy range. If the expected top quark were being produced the value of R would rise by $\frac{4}{3}$ (three colours times the value of the top charge squared). Despite this annoying absence, PETRA has provided a wealth of information on a variety of other phenomena.

During our discussion of the basic reactions of $e^+ e^-$ collisions, we looked at single-photon emission processes only. In fact double-photon exchange is also possible, see Figure 37.2. Although smaller at low energies, say below 3 GeV, it becomes more important as the energy gets larger until at 30 GeV it dominates the collisions.

By studying reactions such as $e^+ e^- \rightarrow e^+ e^- +$ hadrons, we are in fact looking at the material final states of photon–photon collisions. These are currently under study at PETRA.

Although these photon–photon reactions are interesting, the most significant results have extended our knowledge of the simpler one-photon

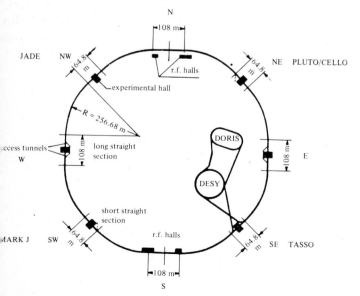

annihilation process. We have said in the past that the ratio R is a constant given by the sum of the squares of the quark charges, but this is in fact only the simplest approximation to its full behaviour in QCD. In the full theory of quarks and gluons, there is a possibility of one of the outgoing quark–antiquark pair radiating a gluon and this so-called radiative correction spoils the constancy of R. Figure 37.3 shows the process concerned and the effect it has on the prediction for R. The experimental measurements are perfectly compatible with the modification predicted by QCD but are by no means conclusive.

Fig. 37.2. Two-photon exchange processes in e^+e^- collisions, can give rise either to an additional e^+e^- pair (*a*), or a shower of hadrons h (*b*).

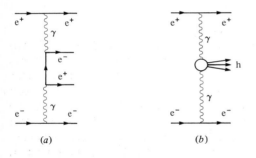

(a)   (b)

Fig. 37.3. Gluon radiative corrections to $q\bar{q}$ production and the variation in R predicted as a result by QCD.

37.2 Jets

At PETRA energies, new evidence becomes available to support the reality of the quark and gluon sub-processes which we believe mediate the process $e^+ e^- \rightarrow$ hadrons. This evidence is in the form of jets of hadrons which are the observable trails of quarks and gluons. In many kinds of elementary particle reaction, the hadrons produced seem always to possess a limited momentum perpendicular to the momentum of the state preceding their creation. Also, the number produced always increases rather slowly with energy. In $e^+ e^-$ reactions this means that, as the energy of the annihilation is increased, the hadrons will emerge in increasingly collimated jets along the opposite directions of the outgoing quarks. At the highest energies these jets are obvious from a visual inspection of the outgoing hadron tracks, see Figure 37.4(*a*).

One of the most important properties about these jets is that their axes preserve the directions of motion of the outgoing quarks. This is significant because the directions of emergence of the quarks, from the point of the reaction, depend upon their spin. Thus the measurement of the angular distribution of the axes of the outgoing hadron jets indicates the spin of the quarks. If we assume the quarks to be spin $\frac{1}{2}$, then the angular distribution of their frequency of emergence can be shown to be proportional to $(1 + \cos^2\theta)$, where θ is the angle between the incoming beam axis and the outgoing jet. Other quark spins would give rise to different angular dependencies. It is possible to infer the jet axis from the observed hadron tracks by performing a

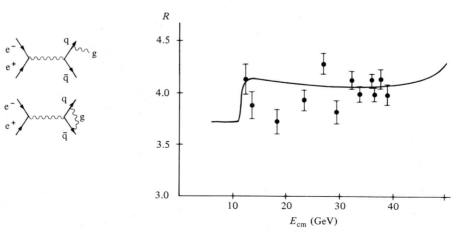

numerical computer search for the direction, amongst all those possible, with respect to which the sum of the transverse momenta of the hadrons is a minimum. Figure 37.4(*b*) plots the angular distribution of these inferred jet axes. We can clearly see that the distribution obtained confirms our assignment of spin $\frac{1}{2}$ for the quarks.

As we have mentioned, more-complex processes are predicted in QCD. In particular, if the energy is high enough, it is possible for a gluon radiated from one of the outgoing quark–antiquark pair to form a separate jet of its own giving rise to a three-jet event, see Figure 37.5. These have also been found in PETRA and are currently quoted as the firmest dynamical evidence available for the existence of gluons.

Another sort of three-jet event also supports the existence of gluons. We have seen previously that the only way in which the psi meson can decay is via three gluons. Because the mass of the psi is comparatively small there is little hope of discerning any jet structure. But the massive upsilon meson must also decay via three gluons and detection of the three resultant gluon jets is a far more feasible proposition. These jets are not as obvious as the true high-energy jets discussed previously, but a three-jet structure can in fact, be perceived in the anisotropy of hadron emission about the interaction region, see Figure 37.6.

37.3 The weak force in e$^+$ e$^-$

In all the processes described above, it was assumed that the electron–positron pair annihilates only to a virtual photon. In fact, another option is that the pair may annihilate into a virtual Z^0 boson and that this weak interaction probe will then be able to probe the 'weak' content of the vacuum. But because the real Z^0 boson is very massive (estimated at approximately 90 GeV), the Z^0 produced in

Fig. 37.4. (*a*) An obvious two-jet event in high-energy e$^+$e$^-$ annihilation, resulting from the emergence of a quark–antiquark pair. (*b*) The angular distribution of jet axes emerging from e$^+$e$^-$ collisions follows the $(1+\cos^2\theta)$ law expected from the production of intermediate spin-$\frac{1}{2}$ quarks.

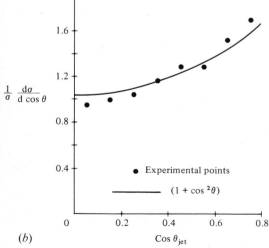

(a)

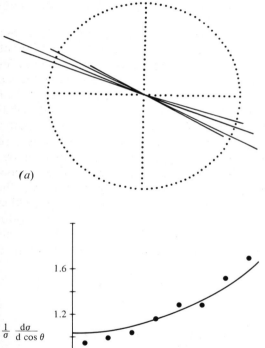

(b)

$\frac{1}{\sigma}\frac{d\sigma}{d\cos\theta}$

• Experimental points

——— $(1 + \cos^2\theta)$

Cos θ_{jet}

Fig. 37.5. A three-jet event in e$^+$e$^-$ annihilation resulting from a quark–antiquark pair plus a radiated gluon.

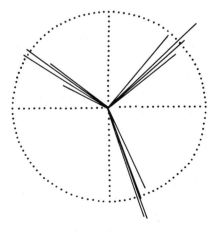

PETRA (maximum energy 38 GeV) will be very virtual (or 'far off mass-shell', to use the technical phase). This means that its effects will be very diluted by the predominant photon annihilation channel. However, as the energy of the electron–positron annihilation is made higher, the photon is removed further off its mass shell ($E = pc$ is more seriously violated), whilst the Z^0 boson is moved closer to its mass-shell, ($E^2 = p^2c^2 + m_0^2c^4$ is less seriously violated). This means that the weak interaction effects of the Z^0 boson become increasingly important compared to the electromagnetic effects of the photon. When (in future years) the energy of the electron–positron annihilation equals

M_{Z^0}, a real Z^0 will be produced at rest, at which point weak interaction effects will be the dominant force shaping the gross features of the final state.

At the highest energies now available in PETRA, it is just possible to discern the effects of the weak force. As we have seen, the primary distinguishing feature of the weak force is its parity-violating properties. For the process $e^+ e^- \rightarrow \mu^+ \mu^-$, this means that the outgoing $\mu^+ \mu^-$ pair is not distributed symmetrically about the interaction region as would be the case if only the electromagnetic force was present. Slightly more muons of a given charge sign should emerge into a given hemisphere surrounding the interaction region into the opposing hemisphere, see Figure 37.7.

After intensive searches for this effect it was observed in PETRA in 1981 at the level predicted by the Glashow–Weinberg–Salam model, thus providing a new source of support for the standard model of the electroweak force.

37.4 LEP tomorrow

Without doubt the most convincing demonstration of the correctness of our modern gauge theories of the fundamental forces has been the observation of the $W^{\pm}$ and Z^0 bosons at the CERN p $\bar{\text{p}}$ collider experiment (see Part 9). A more pedantic confirmation of the details of the Glashow–Weinberg–Salam model would be the observation of the real Higgs boson, which is expected to exist as a result of the introduction of spontaneous symmetry breaking (to explain the mass of the $W^{\pm}$ and Z^0). Finally, observation of a new vector meson in $e^+ e^-$ annihilation should provide evidence for the existence of the sixth top quark to complete the roll-call of three generations of fundamental particles.

Discovery of all of these particles would indeed be a convincing triumph for the standard gospel of particle theory, as explained in the preceding chapters. To achieve this, plans are afoot

Fig. 37.6. Evidence for three-jet anisotropy in the emission of hadrons from upsilon decay. Sphericity is a function of hadrons' momenta reflecting their emission isotropy.

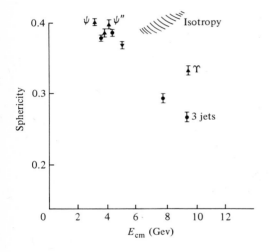

Fig. 37.7. The annihilation of the e^+e^- pair into a Z^0 boson gives rise to an observed asymmetry in the distribution of similarly charged muons emitted.

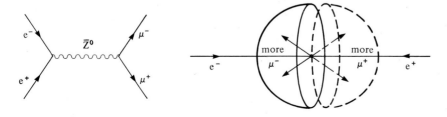

for the construction of a huge new accelerator with an electron–positron collision energy of 270 GeV, which should be amply sufficient to produce annihilation into a single Z^0 and the subsequent production of a $W^+ W^-$ pair. The modestly named Large Electron Positron ring (LEP) is planned for construction at the CERN laboratory at Geneva with a 31 km circumference, much of it underground beneath the Jura mountains, see Figure 37.8. With a planned construction cost of $1 billion (American) and an operating power approaching 100 megawatts (equal to a substantial fraction of Western

Europe's total power consumption), construction of this machine has already been started with a planned commissioning date in 1987.

Other more speculative targets for the particle hunts at LEP will include, naturally, the quarks themselves whose emergence from the SPEAR ring at SLAC was one of the major reasons for its construction. Similarly, magnetic monopoles first predicted by Dirac in 1934 are always the subject of searches in new machines. Finally, there will be the possibility of detecting any of an array of particles which are predicted by the more speculative techni-colour and super-symmetric theories mentioned in Part 9, discovery of any of which would certainly be a turn-up for the books (particularly this one).

Fig. 37.8. A schematic view of the LEP accelerator to be constructed by CERN for very high-energy e⁺e⁻ experiments.

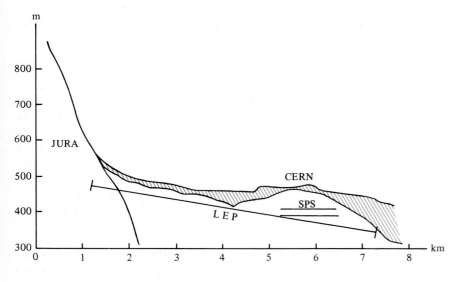

Part 9
Research in progress

The hunt for the $W^{\pm}$, Z^0 bosons

38.1 Introduction

These intermediate vector bosons of the weak force have been by far the most eagerly awaited particles of recent years, for their existence is crucial to the validity of the Glashow–Weinberg–Salam theory of the weak force and, by implication, to the acceptability of all the modern theories of the 'gauge' type. Not surprising then that between 1976 and the present day an increasing experimental effort has been dedicated to their detection. But, as we saw in Chapter 24, the masses of these particles are enormous. So just how is it possible to detect them with present-generation accelerators?

The Z^0 bosons are expected to decay into both quark–antiquark pairs and lepton–antilepton pairs. So, by the microscopic reversibility of particle reactions, we may anticipate that sufficiently energetic collisions between quarks and antiquarks, or leptons and antileptons, will produce the Z^0 boson. As we have mentioned previously, the 'clearer' of the possibilities is doubtless provided by lepton–antilepton annihilations, such as the e^+e^- experiments described in Part 8. The LEP accelerator planned for CERN by the late 1980s is justified largely by this prospect despite the intrinsic difficulties of handling very high-energy e^+e^- beams (rapid energy loss due to synchrotron radiation requiring large power consumption and large-circumference storage rings).

A more available and less-expensive means of discovering the $W^{\pm}$, Z^0 bosons (although a less-satisfactory environment in which to study them) is provided by using the quark–antiquark collisions occurring during proton–antiproton annihilations – and accepting the somewhat more-messy final states resulting from the spectator quarks. In such reactions, one of the quarks inside the proton (uud) can annihilate with one of the antiquarks inside the antiproton ($\bar{u}\bar{u}\bar{d}$) to produce the W$^±$ boson if the pair is dissimilar (e.g. u$\bar{d}$ or $\bar{u}$d), or the Z^0 boson if the pair is similar (e.g. u$\bar{u}$ or d$\bar{d}$), see Figure 38.1.

In 1976, Carlo Rubbia and David Cline and their colleagues suggested converting a conventional, 'fixed target' proton accelerator into a proton–antiproton machine to allow the earliest possible prospect for viewing the bosons.

Fig. 38.1. The mechanisms for W$^±$, Z^0 boson production in pp collisions. In all cases h indicates the hadronic debris resulting from the presence of the spectator quarks.

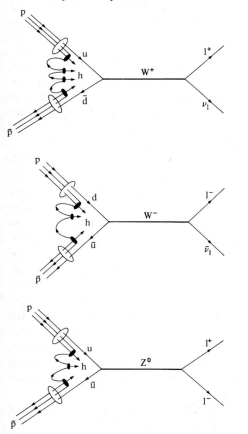

38.2 The CERN p$\bar{p}$ collider experiment

Following the general acceptance of the p$\bar{p}$ idea, the accelerator chosen for the job was the super-proton synchrotron accelerator (SPS) at CERN which, as of 1976, was one of the highest-energy machines in the world, able to accelerate protons to 400 GeV. The great beauty of the p$\bar{p}$ idea is that, because the antiparticles ($\bar{p}$) have the opposite charge to, but the same mass as, the particles (p), the accelerator configuration which accelerates protons in one direction will automatically accelerate antiprotons in the opposite direction. So the one-beam ring of the original SPS was made to accommodate the two counter-rotating beams of p and $\bar{p}$. In the process of this conversion, each beam was designed for an energy of 270 GeV, giving a head-on collision energy of 540 GeV. This, of course, is far greater (over five times) than the thresholds for W$^±$, Z^0 production, but this is necessary as only a fraction of the energy will go into the quark–antiquark annihilations. The rest stays with the spectator quarks and gives rise to complicated, long-range hadron production.

Although the basic idea behind the experiment is simple, the reality is complicated by the absence of naturally occurring antiprotons. These must be painstakingly manufactured in preparatory particle collisions and stored until a sufficient number have been collected to form a beam of sufficient density (referred to as luminosity) for an observable rate of reactions to be possible. Achievement of this 'antimatter factory' is one of the wonders of modern physics and demonstrates convincingly the sophistication with which the experimentalists are now able to control elementary particle beams, through a veritable 'spaghettiscape' of accelerators. See Figure 38.2.

Initially, protons are accelerated to 26 GeV in the 1959 proton synchrotron (PS) machine and are collided into a tungsten target. From the input of 10^{13} protons, approximately 20 million antiprotons of about 4 GeV energy emerge. These are then piped to the antiproton accumulator where an increasing number of such collision bunches are stored. When a bunch first enters the accumulator it is regulated or 'cooled' to remove all random components of the antiprotons' individual movements. This is achieved with a sophisticated control system which detects any such deviation of the antiprotons' movements from the ideal orbit, flashes a signal across the diameter of the accumulator ring and 'kicks' the antiproton bunch back into shape by the application of a tailored magnetic pulse. All this in the time that it takes the antiproton bunch, travelling at practically the speed of light, to travel half-way around the ring! After about two seconds, the antiproton bunch is sufficiently cooled for it to be manoeuvered by magnetic fields into a 'holding' orbit in the accumulator whilst the next bunch is introduced and cooled, after which it too is

Fig. 38.2. High-energy plumbing. Both the proton synchrotron (PS) and the antiproton accumulator (AA) form an integral part of the CERN SPS pp̄ collider experiment. The intersecting storage rings (ISR) conduct separate experiments.

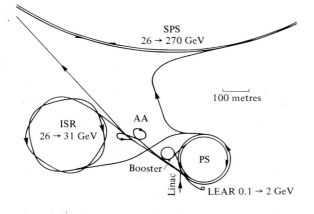

Fig. 38.3. The predicted production rate of W$^\pm$, Z^0 bosons as a function of collision energy, assuming the planned luminosity of the CERN pp̄ experiment.

manoeuvered to join the holding orbit. Over some two days about 60 000 injections are achieved to form a few orbiting bunches of about 10^{12} antiprotons in each. Eventually, the antiproton bunches are sent back to the PS to be accelerated up to 26 GeV, after which they are injected into the SPS. The PS also injects bunches of 26 GeV protons into the SPS in the opposite direction. The SPS can then accelerate the counter-rotating p and p̄ bunches each to 270 GeV when they can be collided through each other at specified interaction regions around the SPS, where various experiments observe the interactions. After this, the same pp̄ beam bunches can go on providing interactions over many hours.

Knowing the luminosities of the beam bunches (equivalent to approximately 10^{30} antiprotons per square cm per second) and using the Glashow–Weinberg–Salam theory to calculate the probabilities of occurrence of the reactions producing the W$^\pm$, Z^0 boson allows us to estimate the probable rate of production of the bosons in the CERN pp̄ collisions. As we can see in Figure 38.3, several hundred are expected each day. The problem then becomes one of finding the bosons amongst the debris of the collisions.

38.3 Detecting the bosons

Having been produced, the W$^\pm$, Z^0 bosons are far too short lived to leave detectable tracks. As is often the case, their presence must be inferred from the behaviour of their decay products. The most significant of these for the W$^\pm$, Z^0 bosons are the charged leptons arising from the decays:

$$W^+ \rightarrow \mu^+ + \nu_\mu,$$
$$W^- \rightarrow \mu^- + \bar{\nu}_\mu,$$

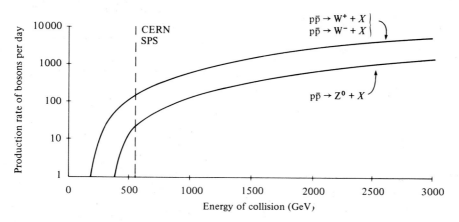

and

$$Z^0 \rightarrow \mu^+ + \mu^-.$$

Several features of the charged lepton distributions emerging from p$\bar{\text{p}}$ collisions can provide the tell-tale signs of the W$^\pm$, Z^0 bosons.

Firstly, the presence of the parity-violating effect will indicate the action of the weak force. The effect of angular momentum conservation and the unique-handedness of the neutrino and antineutrino requires an excess of positively charged leptons emerging in the direction of the incoming antiproton beam (and, similarly, an excess of negatively charged leptons in the direction of the incoming proton beam). This is the same effect as observed in the e$^+$e$^-$ collisions at 38 GeV in the PETRA ring. Although strong circumstantial evidence for the bosons, strictly speaking it indicates only the presence of the weak force and not specifically the method of its propagation via the W$^\pm$, Z^0 bosons.

This more demanding information is indicated by the momenta and energies of the emerging leptons. The production of the W$^\pm$, Z^0 bosons will give rise to a far higher proportion of leptons carrying a significant momentum perpendicular to the axis of the p$\bar{\text{p}}$ collision. By measuring the distribution of this transverse momentum of the emerging leptons, the presence of the bosons should be manifestly obvious, see Figure 38.4(a). Also, the decay of the Z^0 boson involves no 'invisible' neutrinos to carry off any of the energy and so should give rise to the additional distinguishing feature of a very sharp peak in the mass distribution of emerging lepton–antilepton pairs centred on the mass of the Z^0, see Figure 38.4(b).

En passant we should also note that there is a fair chance of detecting the Higgs particle in these p$\bar{\text{p}}$ collisions by similar means. Although less understood theoretically, the Higgs may be produced by the fusion of two gluons from the colliding p$\bar{\text{p}}$ pair, see Figure 38.5(a), or by being radiated off from a Z^0 resulting from q$\bar{\text{q}}$ annihilations, Figure 38.5(b).

38.4 Have they been seen yet?

As if to illustrate the nature of modern discoveries in particle physics, we can report that the W$^\pm$, Z^0 bosons are *in the process* of being

Fig. 38.4. The anticipated effects of W$^\pm$, Z^0 bosons. In (a) we see the increased distribution of leptons emerging with high transverse momentum, p_T. In (b) the mass spectrum of charged lepton–antilepton pairs shows a peak at the mass of the Z^0.

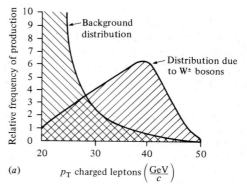

(a)

Fig. 38.5. The production of a Higgs particle in proton–antiproton collisions. In (a) two gluons radiated from the quarks and antiquarks inside the colliding particles fuse together to produce the Higgs. In (b) a Higgs is radiated from a Z^0 boson.

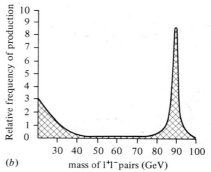

(b)

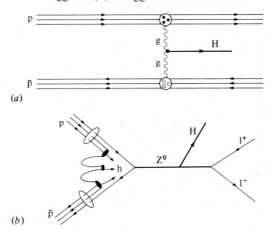

Fig. 38.6. (*a*) The 2000 ton UA1 detector at the
CERN proton–antiproton collider experiment.
(*b*) Tracks emerging from a p$\bar{\text{p}}$ collision as
recorded by the UA1 detector. (Photo courtesy
CERN.)

(*a*)

discovered during the production of this book. At the time of completion of the final manuscript they were expected, during preliminary editing candidate $W^{\pm}$ bosons were announced and just prior to printing, the first Z^0-boson event has been tentatively identified. Thus these much sought-after bosons have been conjured into existence between January and June 1983.

During the latter part of 1982, the SPS $p\bar{p}$ collider achieved a sufficient luminosity to make feasible the discovery of the bosons at the rate of a few events per day. So two separate experiments using different detectors set out to find the bosons. The experiments are referred to as UA1 and UA2, respectively, denoting the different underground areas in which their detectors are located round the SPS ring. Each of the experiment's detectors are very massive assemblies of a variety of particle detection devices (2000 tons in the case of the UA1, 200 tons in UA2). The output of these detectors can be fed directly into on-line computers which allow the subsequent reconstruction and analysis of the tracks of each event (see Figure 38.6(*a*) and (*b*)).

The experiments observed about 10^9 $p\bar{p}$ collisions of which 10^6 were recorded for subsequent analysis. The two experiments then applied various criteria to the observed collisions to find examples of the process,

$$p + \bar{p} \rightarrow W^{\pm} + X$$
$$\hookrightarrow e^{\pm} + \nu.$$

The UA1 experiment sought for two separate classes of event. Firstly, those with an isolated electron with large transverse momentum and, secondly, those events with a large fraction of transverse energy missing (carried off by the undetected neutrinos).

Starting from its initial sample of 140 000 events, the UA1 experimental group was able to identify those events containing the isolated electron with large transverse momentum by applying a series of exclusion conditions (such as demanding that the electron track have originated in the central detector, demanding that other tracks have low transverse momentum, and so on). Eventually, just five isolated electron events were identified.

(b)

Then, starting a new search on a sub-set of 2000 of the original events, the group searched for those events with missing energy (i.e. those events containing energetic neutrinos). This is done essentially by adding up the measured energies of all the observed tracks and checking the total against the known collision energy of 540 GeV. After another set of exclusion conditions just seven events were found, five of which were just those events with the isolated, high-energy electrons. So the UA1 experimental group were able to claim the discovery of five $W^{\pm}$ bosons. Furthermore, by adding up the measured energy of the electron track and the missing energy of the inferred neutrino track, the UA1 group was able to estimate the mass of the $W^{\pm}$ bosons from which these two particles had originated. The best estimate is given as $M_{W^{\pm}} = 81 \pm 5$ GeV in excellent agreement with the prediction of the Glashow–Weinberg–Salam model.

The UA2 experimental group were able to perform a similar analysis of their sample of $p\bar{p}$ interactions, and were able to identify four candidate $W^{\pm}$ boson events and give an estimate of the mass as $M_{W^{\pm}} = 80 \pm^{10}_{6}$ GeV, also in good agreement with the prediction.

These discoveries of the $W^{\pm}$ bosons were announced in January 1983. In May, the UA1 group had collected more $p\bar{p}$ data and were able to offer preliminary evidence for just one Z^0 event in the process,

$$p + \bar{p} \rightarrow Z^0 + X$$
$$\phantom{p + \bar{p} \rightarrow Z^0 +} \hookrightarrow e^+ e^-.$$

Applying yet another series of exclusion criteria, the UA1 experimental group were able to identify just one event in which the emerging $e^+ e^-$ pair emerged back-to-back with equal and opposite high transverse momenta. At the time of writing, the preliminary estimate of the mass of 94 GeV is in good agreement with the prediction.

To be totally confident of these recent discoveries, it will next be necessary to collect many more events containing the $W^{\pm}$, Z^0 bosons and to establish that their properties are as expected (for instance, that their weak interaction couplings have the correct, V–A parity-violating property). If the experimentalists are successful in this, then they will bring to a climax a decade of successful experiments verifying the Glashow–Weinberg–Salam model.

39

Grand unified theories

39.1 Introduction

The success of the Glashow–Weinberg–Salam model of the weak force as demonstrated most recently by the discovery of the $W^{\pm}$, Z^0 bosons, and the evidence supporting QCD, have reduced the scope of particle physics to two sets of particles (the leptons and the quarks) and two forces (the electroweak and the strong), ignoring gravity for the time being. The next logical step is to attempt to formulate the apparent diversity of these particles and forces as the different manifestations of just one type of particle acting through one force. This takes us one step closer to achieving the holy grail of physics, a single theory describing all the forces in nature.

As well as this rather aesthetical motivation, there are also some practical reasons for wishing to formulate a unified theory. Foremost of these is the desire to explain the quantisation of electric charge and why the charge on the electron should be *exactly* opposite to that on the proton (and so an exact multiple of the $\frac{1}{3}$ charges on the quarks). Another reason is to reduce the number of 'free' parameters which must be introduced into the theoretical picture. In the separate theories, about 17 parameters (charges, masses, mixing angles, etc.) must be introduced and their values determined experimentally. In a unified theory many of these will be determined by self-consistency within the theory.

The general approach to the unified theory is

suggested by the success of the principle of gauge invariance applied to both weak and strong forces. Also, the unification of the very dissimilar electromagnetic and weak forces by the use of spontaneous symmetry breaking suggests that this too will play a role in our grand unified theory, commonly abbreviated to GUT. To briefly recap the mechanism: the introduction of Higgs bosons allows the gauge bosons to be given different masses ($M_{W^\pm} \approx 80$ GeV, $M_{Z^0} \approx 90$ GeV, $M_\gamma = 0$). At interaction energies much higher than the gauge boson masses (say 10^4 GeV), they can all be produced easily, meaning that the forces they carry all appear equally important. At lower energy (say about 10^2 GeV), it becomes difficult to produce the $W^\pm$ and Z^0, meaning that the weak force is becoming weaker than the electromagnetic. Finally, at our current energies (up to about 40 GeV), it is very difficult to exchange $W^\pm$, Z^0 bosons and so the effects of the weak force are barely discernible.

39.2 The structure of a GUT

We have seen that the correct gauge bosons for the electroweak force are generated by requiring that the Lagrangian describing the interactions of leptons be invariant under the $SU(2) \times U(1)$ group of transformations. Similarly, the gluon structure of QCD is generated by requiring the quark Lagrangian to be invariant under the $SU(3)_C$ group of transformations. The GUT will be generated by requiring that the total Lagrangian describing the interaction of quarks *and* leptons be invariant under some grand symmetry group which contains $SU(2) \times U(1)$ and $SU(3)_C$ as sub-groups.

There are several possible choices for the grand symmetry group and each gives rise to a distinct particle structure of its own. But the simplest and, arguably, the most successful, is the theory based on $SU(5)$ symmetry. In this theory, the basic entity is a five-component vector containing the right-handed components of each of three colours of the down quark, the positron and the electron antineutrino (which exists only in a right-handed state). Various combinations of this multiplet can reproduce the quantum numbers of all other quarks and leptons. An $SU(5)$ transformation on the basic multiplet thus corresponds to the redefinition of what is a quark and what a lepton. Requiring the Lagrangian to be invariant under the

local group of $SU(5)$ transformations, then, leads to the introduction of a 24-plet of gauge bosons. Of these, 12 are familiar (the photon, $W^\pm$, Z^0 bosons and eight gluons). The remaining 12 are new bosons denoted X; these carry new forces which can transform quarks into leptons and vice versa (see Figure 39.1).

Even at this stage, charge quantisation can be seen as a natural consequence of this $SU(5)$ symmetry. Firstly, the symmetry demands that the average charge of the basic entity be zero and, secondly, that the charge can change only by amounts carried by the gauge bosons (which turn out to be in strict multiples of $\frac{1}{3}e$). The theory thus accommodates the relationship between quark and lepton charges.

Of course, in the observed world we see three known forces but not the new forces transforming quarks into leptons. This is due to the spontaneous breaking of the $SU(5)$ symmetry by the introduction of a suitable configuration of Higgs fields. By this means, the correct masses for the photon and the $W^\pm$, Z^0 bosons are generated, whilst the masses

Fig. 39.1. The $SU(5)$ symmetry of the GUT causes transformations between the members of the basic **5** multiplet (the right-handed components of the three colours of the down quark, d, the positron and the antineutrino). The gauge bosons include the gluons (g) of the QCD $SU(3)_C$ theory and the γ, $W^\pm$, Z^0 bosons of the electroweak theory.

	d_R^{red}	d_R^{green}	d_R^{blue}	e_R^+	$\bar{\nu}_e$
d_R^{red}	g^0, γ, Z^0	g^{r+g}	g^{r+b}	$x^{red}_{-\frac{4}{3}}$	$x^{red}_{-\frac{1}{3}}$
d_R^{green}	g^{g+r}	g^0, γ, Z^0	g^{g+b}	$x^{green}_{-\frac{4}{3}}$	$x^{green}_{-\frac{1}{3}}$
d_R^{blue}	g^{b+r}	g^{b+g}	g, γ, Z^0	$x^{blue}_{-\frac{4}{3}}$	$x^{blue}_{-\frac{1}{3}}$
e_R^+	$x^{red}_{\frac{4}{3}}$	$x^{green}_{\frac{4}{3}}$	$x^{blue}_{\frac{4}{3}}$	γ, Z^0	W^+
$\bar{\nu}_e$	$x^{red}_{\frac{1}{3}}$	$x^{green}_{\frac{1}{3}}$	$x^{blue}_{\frac{1}{3}}$	W^-	Z^0

of the X bosons turn out to be about 10^{15} GeV. These enormous particles lie many orders of magnitude beyond the energy ranges of conceivable accelerators and, similarly, beyond any known spots of very high-energy density in the universe. So they are unlikely ever to be observed directly. Also, their enormous mass means that the forces they mediate will be very unimportant indeed compared to those forces whose quanta are exchanged freely at the energies of interaction concerned.

The grand unified theory therefore incorporates a hierarchy of spontaneous symmetry breaking which reflects the transition from very high-energy simplicity to the low-energy complexity we observe today. At energies well above 10^{15} GeV, all gauge bosons (including the Xs) can be produced freely and all forces are apparent; quarks transform into leptons as easily as they change colours. At about 10^{15} GeV energy, the $SU(5)$ symmetry breaks down to separate $SU(3)$ and $SU(2) \times U(1)$ symmetries and the grand unified force separates into the strong colour force and the electroweak force, whilst the 'new' quark–lepton transforming force becomes unimportant. For energies between 10^4 and 10^{15} GeV, the strong and electroweak forces exist with little interaction between quark and lepton sectors. At about 10^2 GeV, the $SU(2) \times U(1)$ symmetry becomes broken, reflecting the separation of the electroweak force into the separate weak and electromagnetic forces we see today.

This picture of the unification of forces also incorporates the variation in the strengths of charges, depending on the distance from which they are viewed; from the unique strength of the single unified force experienced at energies above 10^{15} GeV, to the three different strengths associated with the three separate forces observed below 10^2 GeV.

The discovery of asymptotic freedom in QCD first pointed up the potential importance of this effect. The colour charge of a quark is spread out on the sea of virtual quark–antiquark pairs and gluons surrounding the quark. So the closer the quark is approached (the higher energy with which it is probed), the less the colour change seems to be. A similar, but less-pronounced, weakening affects the strength of the weak charge on a lepton, whilst, as we mentioned in Chapter 32, the Abelian nature of electromagnetism gives rise to an increase in the effective strength of the electric charge as it is approached.

The relative strengths of the forces known in the current experimental regime (shown in Figure 5.1) and the forms of the variations, as expressed above, lead to the supposition that there is some distance at which all three forces have the same coupling. It is possible to calculate this distance in $SU(5)$ theory and the answer is found to be the remarkably small distance of 10^{-29} cm. The energy required to probe this distance is about 10^{15} GeV, which is just that energy required to enable easy participation of the X bosons. The variation in the strengths of the forces with distance is shown in Figure 39.2.

Other quantities may also be calculated in $SU(5)$ theory, which are far more practical indicators to the success of the theory. Firstly, the ratio of the $U(1)$ coupling constant to the $SU(2)$ coupling constant in the electroweak theory (the weak angle) can be calculated and is predicted to have the value $\sin^2 \theta_W = .20$ compared with the experimental value of $.20 \pm .03$. Secondly, the ratio of the mass of the bottom quark to that of the tau lepton can be calculated to be about 3, compared with the experimental value of about 2.5.

These calculations form the basis for our

Fig. 39.2. Charge shielding and antishielding effects in quantum theory predict the equality of coupling constants (force strengths) at very high energies, about 10^{15} GeV (or very small distances, about 10^{-29} cm). The discontinuity at 10^{-16} cm is due to the dissociation of the unified electroweak force into electromagnetism and the weak nuclear force (which does not act over distances greater than 10^{-16} cm).

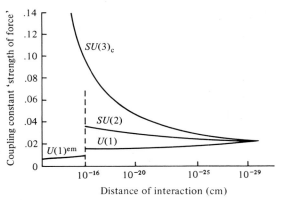

confidence in the grand unified approach. But the validity of the approach hinges on experimental detection of one or more of the unique consequences of the theory, yet to be observed.

39.3 Consequences of grand unification

The most obvious consequence of the standard $SU(5)$ GUT, as described above, is the prediction that baryon number is no longer necessarily conserved in quark–lepton transformations. This allows the possibility of the decay of the proton, which had always previously been regarded as absolutely stable as a free particle, being the least massive of the baryons. Proton decay can occur when a u quark emits an X boson to transform into a ū quark, the X boson being absorbed by the d quark which transforms to a positron (see Figure 39.3). In this way, both the proton and the neutron can decay to a pion and a positron,

$$p \rightarrow \pi^0 + e^+,$$
$$n \rightarrow \pi^- + e^+.$$

If we then go on to note that the neutral pion decays into two photons, and that the positron may

Fig. 39.3. The decay of a proton into a positron and a pion proceeds by the exchange of a superheavy X boson between u and d quarks.

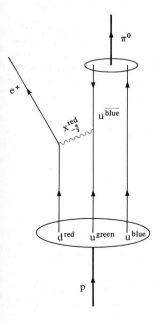

annihilate with any nearby electron again into photons, it leads to the apocalyptic conclusion that the eventual fate of matter will be to transform gradually into radiation acting like some cosmic torchlight battery.

However, the fact that none of us burn up like comets on our daily round gives us some comfort. It is a *very* slow process. It is possible to calculate the decay rate of the proton in the GUT and this provides a value for the mean lifetime of the proton of about 10^{31} years. Bearing in mind the age of the observable universe (i.e. since the postulated Big Bang) is a mere 10^{10} years, the proton is comfortably long lived. But, despite this fantastically long life, it is not beyond the bounds of possibility that we will see (or indeed, have seen already!) instances of proton decay. There are about 5×10^{31} protons and neutrons in 100 tons of matter and so we might expect about five instances of proton decay per year within such a block of matter.

In recent years, several experiments have been mounted in attempts to detect such decays. Generally, the experiments occur deep underground in disused mines in an attempt to filter out all spurious events resulting from cosmic rays, and consist in monitoring reactions occurring spontaneously within a volume of matter. One such experiment, conducted in a gold mine in the Kolar goldfields near Bangalore in India, has claimed six events as candidates for proton decay, which together imply a proton lifetime of about 6×10^{30} years. What is more, three candidates show no tracks intercepting the sides of the volume of matter (iron, in this case), which strongly suggests that the events occur spontaneously within the volume of iron – the expected background from cosmic neutrino interactions with the iron being significantly lower (less than one event expected over the duration of the observation). However, two US experiments have failed to detect any such decays and these place a lower limit on the proton lifetime of about 2×10^{30} years, coming perilously close to denying the candidates from India. Within the next year, similar experiments will commence in several mines in the USA, as well as in tunnels under the Alps. These experiments should detect the anticipated decays if the proton lifetime is less than 10^{33} years, which is certainly the prediction of the standard $SU(5)$ GUT. In the event of no decays being found, then some revision of the exact form of GUT applicable to the real world may be necessary.

Another consequence of GUT is the possibility that neutrinos may, after all, possess a small mass and that, as a consequence, neutrinos may oscillate between one type (electron-, muon-, tau-type) and another, thus violating the previously sacrosanct law of lepton-type number conservation. In the Glashow–Weinberg–Salam theory of the electroweak force, the unique-handedness of the neutrinos and the particular form of Higgs fields introduced to accomplish the spontaneous symmetry breaking ensure that the neutrinos have no mass. This is true also in the simplest $SU(5)$ GUT. However, it is possible to formulate acceptable GUTs either with a more elaborate configuration of Higgs fields and/or with a larger set of basic particles (such as a right-handed neutrino, for instance) which result in a very small mass for the neutrino of between 10^{-3} and 10 eV.

When such a mass is present, an effect may occur similar to the $K^0 \bar{K}^0$ mixing we discussed in Chapter 15. That is, the particular mass states of the neutrinos are not identical to the eigenstates of the weak force. This leads to the oscillation of the type of the neutrino beam as it propagates through space, just as the K^0 beam in Chapter 15 was seen to transform spontaneously into $\bar{K}^0$.

Because of this effect, one would expect a beam of only electron-type antineutrinos (such as emerge from a nuclear reactor) to oscillate into muon-type antineutrinos and thereby give rise to a decrease in the frequency with which electron-type reactions $\bar{v}_e + p \rightarrow e^+ + n$ can be observed in the vicinity of the reactor (from rates expected in the non-oscillating neutrino theory). Several experiments are currently looking for neutrino oscillations in experiments based around nuclear reactors and at the time of writing, the experiments are in disagreement as to whether or not the effect actually exists. One group at the University of California claim to have observed the effect, another based in Grenoble in France deny it.

Although neutrino oscillations imply the existence of a finite mass for the neutrino, the converse is not necessarily true. The neutrino may have a finite mass without giving rise to oscillations. Other types of experiment can attempt to observe the neutrino mass more directly by very accurate measurements of the masses of all the other products of nuclear β decay. One recent Russian experiment has reported a finite mass for the neutrino but this has not yet been corroborated elsewhere. The possible existence of a very small mass of the neutrino may sound at first to be of rather marginal interest but, in fact, the cosmological consequences of such a discovery are enormous.

The eventual fate of the universe is either for it to go on expanding, as it is at the moment, indefinitely, or for the expansion to slow down and eventually stop and for the universe to collapse back down into a universal black hole in a kind of reverse film of the original Big Bang. Which of the two possibilities comes to pass depends most importantly on the matter density in the universe. Current observations and assumptions are rather inconclusive as to which of these two fates may eventually occur. But with a finite mass, almost however small, for the neutrinos, their abundance in the universe would, without doubt, add significantly to the universal matter density and ensure the gravitational collapse of the universe.

As we are about to see, this is only the beginning of a series of connections between particle physics and cosmology.

39.4 Grand unification and cosmology

Already, our present-generation particle accelerators are examining phenomena at such high energies (540 GeV at the CERN p$\bar{\text{p}}$ collider) that there are practically no known parallels elsewhere in the universe. So our theoretical speculations concerning reactions with energies of order 10^{15} GeV would seem to be somewhat superfluous. But, there is one laboratory in which such reactions were commonplace, and that is the laboratory of the early universe *very* soon after the initial Big Bang. Our accelerator experiments can be thought of a probes back in time (on which scale the 540 GeV at CERN corresponds to a time 10^{-10} s after the Big Bang) and the GUTs can be viewed as our attempts to write the history of the very early universe.

There are also some practical points to the connection. For we may be able to use cosmological observations to complement the expensive and complicated accelerator experiments of high-energy physics and, conversely, we may be able to explain the state of the universe on the basis of our knowledge of particle physics. In this brief review, we will look at one topic in each category. Firstly, there is the possibility that cosmological observations may determine the number of neutrino types that can exist, and so, if you believe in the generation structure of the particles, in the total fermion population of the universe. Secondly, we will look at the possible explanation of why the universe is made of matter and not antimatter.

39.4.1 *Neutrino-type number determination*

The observation which may provide this apocalyptical estimate is the abundance of cosmological helium – currently estimated to form about 27% of the mass of the universe. The importance of helium lies in the fact that stellar processes can account for only a very small fraction of this figure. The majority is thought to have been formed by nucleosynthesis about 14 s after the bang when the temperature was about 10^9 K, the actual amount formed depending crucially on the neutron/proton ratio existing at that time. Conventional Big Bang theory allows this ratio to be calculated assuming that equal numbers of neutrons and protons existed in some initial equilibrium state of 10^{11} K at about 10^{-1} seconds after zero time. Two main factors influence the evolution of the neutron/proton ratio. Firstly, various elementary particle reactions closely related to nuclear β decay transform neutrons into protons and vice versa. Secondly, the expansion rate of the universe governs which of the reactions is dominant at any time and so determines the changes in the ratio. It so happens that the expansion rate depends on the number of neutrino types present and so, indirectly, this number determines the abundance of helium in the universe. The current indication is that no more than four different types can be accommodated within the observed helium abundance, signifying the existence of only one more generation of particles beyond those for which we already have evidence.

39.4.2 *Matter–antimatter asymmetry*

It now seems likely that there are no large concentrations of antimatter in the universe, and almost certainly none within our local cluster of galaxies. The question arises naturally of how this has come about, assuming (the only philosophically attractive option) that in the instant of the Big Bang

no divine guiding hand arbitrarily decided on one or another (i.e. assuming the initial state to be matter–antimatter symmetric). Although still very speculative, a convincing chain of reasoning can be advanced.

At a time less than 10^{-35} s after the Big Bang, the temperature of the fledgling universe would have been about 10^{28} K, corresponding to an average energy of the material particles of about 10^{15} GeV. In this regime the super-heavy gauge bosons X and their antiparticles $\bar{X}$ could be produced with ease in particle collisions and the total equal populations of X and $\bar{X}$ could remain in equilibrium (i.e. as many would be produced in collisions as would decay).

But, as the universe expanded and cooled, the X and $\bar{X}$ could no longer have been produced and the numbers of X and $\bar{X}$ present at 10^{-35} seconds would begin to decay. Because of the presence of the **CP**-violating effect described in Chapter 15, there would be no guarantee that the average value of the baryon number of the states into which the Xs decayed would be exactly opposite to the value of the baryon number of the states into which the $\bar{X}$s decayed.

So the **CP**-violating decays of the X $\bar{X}$ could give rise to a net baryon number for the universe from a state at about 10^{-35} s, consisting of an equal number of X and $\bar{X}$. As the average energy of the particles in the universe would then have continued to fall, baryon-number violating processes would have become increasingly insignificant and the net baryon number would thus have become 'frozen in'.

Using the GUTs, it is possible to calculate the net baryon excess generated. This is generally stated as the ratio of the density of baryons N_B to the density of cosmological Big Bang photons (approximately 400 cm^{-3}). The observed value,

$$\frac{N_B}{N_\gamma} \approx 10^{-9\pm1}$$

can generally be reproduced in most GUTs.

Although we have examined appealingly simple pictures of these two topics, we should realise that the application of GUTs to the early universe is highly speculative. We have simply ignored many of the possible unknowns. Amongst these, for instance, are the roles which may have been played by the Higgs particles (and the Grand Higgs necessary in GUTs), the effects of black holes and the effects of magnetic monopoles (predicted by some gauge theories). Whilst the presence of these entities do not necessarily preclude the simple pictures given above, they suggest that they may be at best some approximations to a more complicated truth.

40

The latest ideas

40.1 Introduction

It is a philosophical marvel to watch the progress of knowledge, and nowhere more so than in physics. Man-made ideas are put forward as theoretical speculations for experimental evaluation. Over a period of time, the experiments test the speculations and, if they are 'correct', the speculations become facts, true of the real physical world and altogether independent of humans.

For instance, such has been the case in the progress of the Glashow–Weinberg–Salam model. In, say, 1970, this model was most distinctly a speculative theory. But now, after more than a decade of repeated experimental verifications, it is generally accepted as the factual account of how the weak interactions work.

In this chapter, we will review some of the modern (and some not so modern) ideas which are still most definitely speculative. Experiments are currently testing all of them, directly or indirectly, and it is a fascinating intellectual game to try at this stage to spot the true 'facts' hidden in the lather of speculation.

40.2 Super-unification

In the last chapter, we saw how the promising GUTs can provide a credible framework for the unification of the electroweak and strong forces. Next, we will look briefly at a class of theories which may describe gravity as well, leaving us with a single unified theory of all the forces in nature. These theories are called super-gravity theories, for reasons which will become obvious. At the time of writing, they are almost entirely hypothetical, with few testable predictions resulting. But they have excited much attention on the grounds of their power, elegance and economy.

We saw, in Chapter 5, some of the difficulties associated with the description of gravity. It has the effect of distorting space–time and there is no calculable quantum theory of the force. So, at first sight, it seems a pretty hopeless candidate for unification with the other forces – which do not affect space–time (can be described in conventional, 4-dimensional 'flat' space–time) and which are well described by quantum-field theory. But all is not lost.

The starting point for the new theories is back with the idea of symmetry as the origin of forces. In some sense, the theory of gravity provides the most obvious example of all, as it deals directly with space and time. The symmetry in question is a simple displacement through space. Obviously, we expect physics and its laws to be the same on the moon as on the earth and so we must be able to displace the coordinate system without changing the results of physics. Such a shift, technically called a 'global Poincaré' transformation, does indeed leave the laws of physics invariant, as we might expect for such a simple manoeuvre. But we can go further and ask that the laws of physics do not change even if we shift each point on the coordinate system by a different amount. Suddenly this seems ridiculous! If the points of a regular grid are each displaced by an arbitary amount, the result is nothing like what we started with – it is chaos. We need something to restore the laws of physics to their original form after such a chaotic transformation, and this turns out to be the gravitational field which 'straightens out' the distorted space–time resulting from this 'local Poincaré' transformation. So, the gravitational field is the gauge field required to maintain the symmetry of physics under local translations.

Any particle can experience the force of gravity by interacting with the gauge particle of the gravitational field, the graviton. As we mentioned in Chapter 5, this is a massless, spin-2 particle; massless because gravity is of infinite range and spin 2 because it is always attractive. But, as we also mentioned in Chapter 5, it is impossible to construct a sensible quantum field theory of gravity from here on as it is impossible to renormalize the theory. The quantum corrections to gravity turn out to be infinitely large, which is what is behind the apparent incompatibility between quantum theory and gravity.

At this point, we can say either of two things. Either the two theories are genuinely contradictory or we are missing out on some vital principle which might reconcile these two great theories. As we know they both work well in their own areas, it seems unfair to say that one is right and one is wrong, so we must look for something new. As we have already seen how important symmetry is, perhaps we might think of looking for some new transformations with which to test the laws of physics.

The new principle was discovered in the early 1970s by several physicists, working independently, in Russia, Europe and America. The symmetry in question is that between particles of different spins (i.e. between fermions and bosons). These had previously been regarded as wholly different entities and so a symmetry linking the two was entirely novel. It has since become known as super-symmetry. The physicists found transformations between fields representing particles of different spins and the challenge was to see if the laws of physics would remain the same under the transformations, and what would happen when the transformations were performed locally (e.g. differently at each point in space).

The first remarkable discovery was that the transformations which change the spin of the particles also shift them in space. So the super-symmetric transformations automatically include the simple Poincaré transformations we mentioned earlier. When we make all the transformations local then we are led to local Poincaré transformations and so to gravity. From examining the spins of the elementary particles it looks as if we derive a theory of gravity!

The other remarkable fact about the transformations is that the laws of physics will still look different even after the gravitational field has been introduced. Simple gravity alone is not enough to ensure super-symmetry. Again, something new is needed. We have to introduce another gauge field to restore symmetry, and the particle it represents turns out to be a massless, spin-$\frac{3}{2}$ oddity – the gravitino. When its field is included, the laws of physics are invariant under local super-symmetry transformations and we have introduced a new theory of gravity.

The new principle of super-symmetry has led us to introduce two force fields. One is just the field of the graviton, which means that we have included the old quantum theory of gravity with all its successes (it will reduce to general relativity when applied to the macroscopic world), but also with its failure – its inconsistency at the quantum level. The second new field represents a spin-$\frac{3}{2}$ particle, totally unlike any of the gauge fields of the other forces. What does it do? We might be worried, because we certainly do not want to introduce any new forces. Luckily, forces corresponding to half-integer spin fields have no classical counterpart, they are a purely quantum effect. This is because they obey the Pauli exclusion principle which stops large numbers of the particles doing exactly the same thing and so prevents any macroscopic effects building up.

We can heave a sigh of relief – the classical theory of general relativity remains unspoilt. So what does this new field do?

We have already seen how the quantum-mechanical probability of an event occurring is found by adding up all the possible exchanges of the gauge fields, so in the theory of super-gravity we have to add all possible gravitino exchanges to the graviton exchanges we already know about. And then we discover something remarkable. In many of the calculations performed so far, the infinite quantities which result from graviton exchange are exactly cancelled by new infinities resulting from the gravitino exchanges and for the first time we achieve an alleviation of the infinities which have plagued quantum gravity for so long.

Unfortunately, it is not the case that all the infinites are banished by super-symmetry. Rather, it has provided only one step (albeit one of great importance) towards a renormalisable quantum theory of gravity. What we need is a general proof that the infinites cancel in all processes, similar to the proofs that exist for the other quantum field theories. Such a proof would be a marvellous thing – the final reconciliation between quantum theory and general relativity.

The formalism of super-gravity obviously has great potential, but what does the theory actually describe? In short, the answer to that is – everything, perhaps! Because we can use the super-symmetry transformations to describe particles of different spins in basically the same fashion, it follows that specification of the transformations will automatically dictate how many particles of each spin can be described. The mathematics of group theory specify which transformations are feasible and these specify the spectrum of the elementary particles. So, starting with the spin-2 graviton, we can use the transformations to find the spin-$\frac{3}{2}$ gravitinos, the spin-1 particles like the photon and the $W^{\pm}$ bosons, the spin-$\frac{1}{2}$ particles such as the electrons and neutrinos and the spinless particles. The complete set of the elementary particles might be generated by the transformations. So all the particles are present and so is gravity.

The other forces between the particles can be introduced by ensuring the invariance of the theory under the local transformations of the group symmetries governing the forces (i.e. $SU(5)$ in the case of the standard GUT). At best then, we have a completely unified theory of all the particles and all the forces we know in nature – the ultimate dream of theoretical physics.

But, technical difficulties remain in implementing this scheme. One major problem is that the spectrum of particles with different spins does not correspond exactly to the known (or anticipated) spectrum of the observed elementary particles. This is perhaps not too serious. There may be many new particles which we have not yet discovered and there may be mechanisms by which the particles of the super-symmetry spectrum can dress themselves up into bound states which then might correspond to the observed particles. Even as it is, the predicted spectrum looks promising. However, the ability of the theory to predict all the particles which should exist is terribly restricting and this will be a severe test of the super-symmetry principle.

Also, mathematical proof of the renormalisability of supergravity is essential to ensure a workable quantum field theory of general relativity. The potential of a completely unified theory of the particles and forces in nature is enormously attractive but, as we might have expected, the problems involved in its complete formulation are correspondingly daunting.

40.3 Technicolour theory

The standard gauge theories have had great quantitative success in describing (and predicting) various experimental measurements (e.g. the g factor of the muon, parity-violating phenomena, the weak angle, etc.). But they are not without their shortcomings and attempts to avoid the problems have led to another class of theories. These attempt to replace the mechanism of spontaneous symmetry breaking by a method of mass generation resulting from the dynamics of new 'technicolour' forces, introduced especially for the purpose.

In the electroweak theory, we have seen how the weak nuclear force is carried between particles by the massive W bosons. These particles have their masses generated by absorbing some components of a hypothetical Higgs field introduced for just this purpose. The remaining components of the Higgs field represent real elementary particles which should be observable in experiments but have yet to be discovered. Unfortunately, if the Higgs particles are truly elementary particles then the mathematical description of *their* masses runs into problems and the best way to avoid these is to drop the idea that the Higgs are elementary and allow the possibility that they are composites of still more-elementary particles. The technicolour theory attempts to describe this possibility and, as its name suggests, it borrows its approach from QCD (colour theory).

Technicolour theory proposes the existence of a family of elementary techni-fermions (particles with a new technicolour charge and half-integer spin) which act as the sources of new technicolour forces. By dynamics presumably similar to the (unknown) dynamics of QCD, the techni-forces bind the techni-fermions into permanent combinations, some techni-mesons being absorbed into the mass of the W bosons, others remaining to be observed as real particles. But in the electroweak theory, the Higgs field also generates the masses of the quarks and the leptons (like the electron) and it is in attempting to provide this service that the technicolour ideas run into difficulties; it would seem to require the existence of a direct four-fermion interaction (two fermions and two techni-fermions), which was long since known to be mathematically unstable. (This was the original reason for the introduction of the intermediate W bosons.) To avoid this, it is necessary to introduce a family of massive techni-bosons to carry the techni-forces between fermions. The techni-bosons must themselves have their masses generated in a fashion similar to that used for the W bosons and so this leads to an iteration of the technicolour scenario – the so-called 'extended technicolour theory'. Although this theory may provide an acceptable view of mass generation, it leaves behind a plethora of techni-mesons which should be observable in accelerator experiments.

It is possible to estimate the properties of these techni-mesons and it is generally predicted that their masses should be below about 10 GeV. These should be produced quite easily in electron–positron annihilation experiments, subsequently preferring to decay into the tau heavy lepton and its neutrino, or into particles containing the heavier charmed or strange quarks. Also the techni-meson might be seen amongst the products of D-meson decays or of bound states of the yet-to-be discovered sixth 'top' quark and its antiquark. Existing data on electron–positron annihilations comes periously close to ruling out the existence of these techni-mesons, and future data, especially from the next generation of electron–positron colliders, may well prove conclusive one way or the other.

40.4 Magnetic monopoles and solitons

In one of the classical papers of theoretical physics, published in 1931, Dirac predicted the existence of magnetic monopoles. (In the preface to this paper, and almost in passing, he finally identified the holes in the sea of negative-energy electrons as antielectrons rather than protons, as had been his original conjecture.) His motivation for writing the paper was the attempt to explain the origin of the quantisation of electric charge. This he could do only by postulating the existence of a magnetic charge g which is related to the familiar electronic charge e by the relation,

$$2\,e\,g = \hbar\,c\,n,$$

where n is an integer.

Although startlingly strange, the idea of a magnetic monopole is theoretically attractive. Immediately, the existence of a source of the magnetic field allows Maxwell's equations describing electromagnetic fields to be written in a symmetric form with both electric and magnetic fields arising from both electric and magnetic charges. The experimental absence of magnetic monopoles was assumed to be due to their mass and to the strength of the force between monopoles preventing their individual observation. The subject continued to attract theoretical attention (albeit at a rather meagre level) and the magnetic monopole gradually became known as 'a well-known, undiscovered object' which was subject to routine searches in the succeeding generations of higher-energy particle accelerators, in cosmic ray experiments and in bulk matter – all to no avail.

Modern interest in the possibility of magnetic monopoles revived suddenly in 1974 with the prediction of such objects on the basis of modern gauge theory. This was made independently by Gerard 't Hooft and by Alexander Polyakov of the Landau Institute in Moscow. They discovered that magnetic monopoles should exist as the so-called 'soliton solutions' to the successful Glashow–Weinberg–Salam gauge theory of the electroweak force.

Soliton solutions are configurations of fields of finite energy which are both localised and stable. They are an entirely general consequence of classical physics and are well known in subjects such as hydrodynamics. For instance, a normal water wave is initially localised and has finite energy. But it is not stable. As it propagates across the surface of the water it spreads out and dies away. This is because of the dispersion of the wavelengths (i.e. different wavelengths will travel through the medium at different velocities). However, in very particular circumstances, the effects of dispersion can be cancelled out by some non-linear effects between the wave and the medium, giving rise to a non-dissipative, 'soliton' wave. In contrast to the normal dissipative water waves, a soliton wave will not disperse but will propagate as a stable disturbance. This is the origin of tidal bores of which that on the River Severn is perhaps the most famous.

The condition for the existence of soliton solutions in field theory is related to the vacuum structure of the theory and is well illustrated by the example of a 2-dimensional rubber sheet which we may think of as our field. The field has two possible vacuum states (states of minimum energy) corresponding to either of the two sides being face up. When only one vacuum is present, the sheet is flat. Any finite energy disturbance will propagate through the sheet as a wave and will eventually disperse, see Figure 40.1. But when two vacua are present, both sides are face up and a knot must join together the different vacuum regions. Although this knot may move about, it can never disperse while the two vacua remain. This then is simple soliton: a localised concentration of a finite amount of energy joining two distinct, but equivalent, vacua.

As is possible to visualise, solitons (twists in the sheets) and antisolitons (twists in the other direction) can collide and annihilate each other giving rise to a normal, dissipative-type wave. It is also possible for solitons to collide and pass through each other, maintaining their shape and energy through the collisions. Thus soliton solutions to classical field theory can exhibit very particle-like behaviour, which we would normally expect to describe using quantum field theory. The added intriguing fact is that solitons are extended objects and so may help us describe, in a natural way, particles of finite size; in contrast to the invariably point-like particle behaviour modelled in quantum field theory.

The modern excitement began with the dis-

Fig. 40.1. A rubber sheet lying flat can have two vacua when either side is lying flat (*a*). When only one is present (*b*), any disturbance will dissipate. When both vacua are present (*c*), a knot must interpolate between them. This knot cannot disperse for topological reasons.

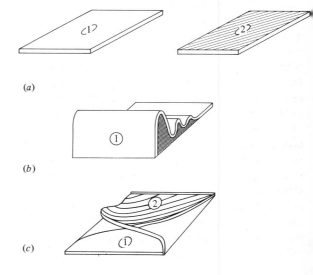

(*a*)

(*b*)

(*c*)

covery of soliton solutions in $SU(2)$ gauge theory. In this case, multiple vacua occur because of the introduction of the Higgs field necessary for spontaneous symmetry breaking. As we saw in Chapter 22, the form of the potential energy of the Higgs field is such that a state of minimum energy exists for any values of the two-component field ϕ which satisfy the equation,

$$\phi_1^2 + \phi_2^2 = R^2.$$

During spontaneous symmetry breaking, the system of fields evolves towards the state of minimum energy. This corresponds to the selection of one particular vector orientation in the space defined by ϕ_1 and ϕ_2. Thus there is a potentially infinite number of the orientations of the vector ϕ in the space defined by ϕ_1 and ϕ_2, which give rise to the state of minimum energy. The solitons in this theory are the kinks in the gauge field which connect the regions of different orientation of the Higgs field. Further work reveals that these particular gauge field kinks contain an isolated unit of magnetic charge and may be extremely massive, say about 10^3 GeV.

Solitons in this theory can also give rise to elaborate effects involving angular momentum. It is possible to show that, in the presence of a soliton, bosons may combine to form a fermion – a previously impossible occurrence. Additionally, it is possible for a soliton to divide a fermion into two half-fermions, also previously impossible.

Magnetic monopoles are also predicted by the GUTs. In these theories, the monopoles are formed on the spontaneous breaking of the grand unified gauge group which, as we have seen, occurs typically at energies around 10^{15} GeV. From this it is possible to show that GUT monopoles must have enormous masses in the region of 10^{16} GeV, or approximately 10^{-8} g. Just as in the case of the grand unified X-gauge particles, there is no possibility of producing such particles in any conceivable accelerator. But also as with the X-particles, it transpires that GUT monopoles should have been produced copiously in the early universe some 10^{-35} seconds after the Big Bang. It is, to say the least, an extremely non-trivial problem for the theories to explain the apparent absence of relic GUT monopoles and their effects. For instance, one plausible effect of their existence is for them to gather at the earth's core where they would most certainly give rise to very different geomagnetic effects than those known.

Moreover, as we have indicated, there are many possible GUTs and super-gravity theories which differ in the symmetry groups governing their Lagrangians and in their particle contents. It is possible that many soliton solutions exist for each of these theories and so give rise to a wide range of entities with a diverse range of properties. This so-called 'non-perturbative structure' of the modern gauge theories presents a whole new area of theory, undeveloped until recently, and holds out some intriguing possibilities. For instance, in some theories objects called dyons are discovered, which carry both electric and magnetic charge. It is conjectured that these dyons could provide some sort of model for the quarks, which would then be bound together by magnetic strings connecting their magnetic charges.

Another leading conjecture is that, in the situation of multiple vacua, a system of fields can effectively undergo quantum-mechanical tunnelling from one vacuum to another (by a mechanism similar to the tunnelling of electrons through potential barriers at the junctions of semi-conductors). We can think of this as the vacuum configuration of the gauge fields flexing into an equivalent but distinct alternative vacuum state. This tunnelling is described by a family of soliton solutions that are transient in time. For this reason they are called instantons. Instantons are not particles and have no direct physical interpretation. Rather, they represent vacuum fluctuations of gauge fields which may lead to some observable forces on nearby particles such as quarks. On the basis of this, some theorists have attempted to explain various meson-mass anomalies as resulting from the effects of instantons on the interquark forces. Others have put forward instantons as yet another mechanism which may provide the understanding of quark confinement. But despite such suggestions, the physical importance of instantons is still far from obvious. Such ideas are currently all very speculative. However, even now it is certain that the modern gauge theories have a rich structure (and, one day possibly, consequences?) far beyond the original reasons for their introduction.

Most recently, a wave of excitement ran

through the physics community at the convincing report of a monopole sighting. In February 1982, Blas Cabrera of Stanford University reported the sighting of a single monopole in an experiment designed specifically for the purpose. The experiment attempts to detect cosmic monopoles incident on the earth's surface. It consists of a small (5 cm diameter) super-conducting ring in which the current is monitored by an extremely sensitive super-conducting quantum interference device (SQUID). Both the ring and the SQUID are then placed in a controlled region of ultra-low magnetic field. In the event of a cosmic monopole transiting through the ring, there will be a long-range electromagnetic interaction between the magnetic charge of the monopole and the macroscopic quantum state of the ring. This will give rise to a distinctive charge in the super-current in the ring corresponding to an integer multiple of the super-conducting flux quantum.

Over a period of 151 days' continuous monitoring of the ring, just one event was discovered which bore the exact hallmarks of a monopole occurrence. Detailed examination of all other possible causes for such an event were unable to reproduce the signal detected and so the monopole assignment has remained a feasible proposition. But, obviously, further sightings must be achieved to establish a discovery and several experiments are currently waiting on the flux of heavenly monopoles.

40.5 Prequarks

QED, QCD and the Glashow–Weinberg–Salam model together form what is referred to as the standard model of the elementary particles and forces. As we have just discussed, the GUTs and super-gravity attempt to progress beyond the standard model by the further unification of the forces involved, whilst assuming the quarks and the leptons to be the fundamental particles. Another route to progress beyond the standard model is the opposite approach; to leave the unification of the forces (at least for the time being) and to attempt to introduce a more fundamental type of particle of which the quarks and leptons are composites. This new type of particle is known generically as a prequark.

It must be said at the outset that there is absolutely no experimental evidence to suggest that quarks or leptons have any internal structure of their own. This implies that any hypothesised sub-structure must be contained within the current experimental resolution of 10^{-18} m. But having accepted this constraint, we can appreciate that there are some powerful motivations encouraging a composite view of quarks and leptons. Firstly, there are an uncomfortably large number of them, for what are supposedly fundamental entities; six flavours times three colours of quark plus six leptons equals 24 quarks and leptons. Secondly, we would like to explain the apparent three-generation structure of the quarks and leptons which introduces more-massive repetitions of the same basic set of quantum numbers. And finally, there is the old puzzle of why the charges on the quarks are exactly $\frac{1}{3}$ multiples of the electronic charge (and thus why the hydrogen atom is exactly neutral).

These motivations are very similar in spirit to the motivations that first persuaded physicists to explain the diversity of atoms in terms of the constituent protons, neutrons and electrons and then, more recently, to explain the multitudinous hadrons as composites of quarks. So the desire to make do with as few basic entities as possible has had remarkable success in explaining several generations of physics. As soon as the proliferation of the quarks and leptons became apparent it was natural to try the traditional approach.

One of the first prequark models to emerge was that of Abdus Salam and Jogesh Pati in 1974. In their model, entities called preons which are characterised by electric charge, colour and a new generation number can combine to form the known quarks and leptons.

Another scheme put forward by Haim Harari of the Weizmann Institute of Science in Israel is the rishon model (where the word rishon comes from the Hebrew word meaning primary). In this scheme, any quark or lepton in any generation is given as a combination of two primal rishons, (the electrically charged T type and the electrically neutral V type). The different generations of quarks and leptons are regarded as the different dynamical excitations of the same system and so do not require the introduction of any new rishons. The charged leptons can consist of only the three charged rishons (TTT) and the neutrinos only of the three neutral rishons

(VVV), but the fractionally charged quarks can be made up of different combinations such as TTV and VVT. The T rishon is put forward as carrying the three colours whilst the V rishon carries the anti-colours. In this way, the fractionally charged quarks can be given colour, whilst the integer-charged leptons remain colourless.

When we combine the rishon content of the three quarks inside the proton with the antirishon content of the electron, then there appear to be equal amounts of matter and antimatter inside the hydrogen atom, in contrast to the usual picture in which (problematically) the universe is dominated by matter (at least locally).

In the rishon model, the quanta of the different force fields such as the photons, the $W^{\pm}$, Z^0 bosons and the gluons are large composites of rishons and antirishons which can alter the rishon content of the particles between which they act, so as to reproduce the observed transformations between particles.

But, though the combinatorics of the pre-quark models allow the quark and lepton spectrum to be built up in a neat fashion, attempts to formulate the dynamics of prequark models soon run into problems. Firstly, it is necessary to propose that some new force holds the prequarks together inside the quarks and leptons. It is easy enough to think of this by analogy with QCD. We can propose that each prequark carries some new kind 'hyper-colour' charge which acts as a source of the new, ultra-strong hypercolour force that binds them together. Also, we can go on to postulate that what we now call the weak interaction is, in fact, just the residual effect (i.e. the van der Waals force) of the shorter-range hypercolour interactions of the pre-quarks (just as we now view the strong nuclear force as the residual effect of the chromodynamic force between the quarks).

Unfortunately, such a seemingly feasible theory alone cannot be enough. By Heisenberg's uncertainty principle we know that any particle confined to a small region can have no less than a certain energy dictated by the uncertainty in its momentum. For a region as small as 10^{-18} m, the minimum energy permissable is of the order of 100 GeV. This means that we might naively expect the simplest composite particles with a sub-structure of this dimension to have a mass also of order 100 GeV, and that the dynamical excitations of the basic composite (corresponding to the higher generations) would have masses approximating to multiples of this amount. This is in contrast to the mass range of the known quarks and leptons of 0–5 GeV. Any prequark model then is only feasible if the enormous (negative) binding energy of the hyper-colour forces miraculously cancels out the kinetic energy of the constituents to leave almost or exactly massless composites.

One current conjecture is that the miraculous cancellations are, in fact, enforced by some symmetry governing the composites' behaviour. This is currently the subject of active research. Until this fundamental problem is alleviated, the idea of prequarks as a way forward will remain somewhat eclipsed by the more popular GUTs.

41

The beginning of the end?

41.1 Introduction

As we noted in the preface, history has an uncanny knack of repeating itself. The last decade has echoed the progress of the 1920s in the way that both theory and experiment have advanced in a complementary fashion to provide a convincing understanding of the microworld at a level several orders of magnitude smaller than could be understood, 50 years ago. Whereas in the 1920s energies of a few eV probed the atomic distances of about 10^{-8} cm, we now use energies of hundreds of GeV to probe distances of 10^{-16} cm.

As a result of our recent successes, we now believe we have a reasonable understanding of the world on this scale. It has been the path to this understanding which we have seen described in this book. Having trod the path, the understanding itself can be summarised meaningfully in just a few short sentences (which is always a good sign!).

We believe that the material particles in the universe consist of elementary fermions. Of these, six leptons $\binom{e}{\nu_e}$ $\binom{\mu}{\nu_\mu}$ $\binom{\tau}{\nu_\tau}$ exist as free, point-like particles, whilst the 18 quarks (six flavours times three colours) $\binom{u}{d}_{rgb}$ $\binom{c}{s}_{rgb}$ $\binom{t}{b}_{rgb}$ seem to exist only in certain combinations, specifically, qqq as the baryons and q$\bar{\text{q}}$ as the mesons. What is more, the six leptons and six flavours of quark both seem to fall in the pattern of three generations, as indicated by the brackets. Each generation is simply a different mass version of the same set of basic quantum numbers. The electromagnetic and the weak and strong

nuclear forces between the quarks and leptons are carried by gauge bosons whose properties are described (or predicted) by the symmetries observed to govern particle interactions. These are the photons of the electromagnetic force, the $W^\pm$, Z^0 bosons of the weak force and the gluons of the strong force.

There is strong evidence for the unification of the electromagnetic and weak nuclear forces as one single electroweak force. This is thought to become apparent when interaction energies are high enough to make the mass differences of the γ, $W^\pm$, Z^0 bosons irrelevant. Also, there is growing support for the grand unification of the electroweak and strong forces at ultra-high energies.

41.2 Where to next?

Having written this credible gospel of the microworld, many physicists are now pondering the long-term future. Assuming that the standard electroweak theory is confirmed by the experiments at the CERN p$\bar{\text{p}}$ collider and LEP, and that the third generation of fermions is satisfactorily completed, there is consternation as to what might, or might not, happen next.

In the standard minimal $SU(5)$ GUT, essentially no new phenomena are anticipated between the energy scales of 100 GeV (where electroweak unification occurs) and 10^{15} GeV (where grand unification occurs). We are now probing the former scale with the enormously complicated and expensive accelerators we have described. At the present time, there seems to be no practical means of approaching the latter and thus the physics of grand unification. Also, there are good reasons for supposing that the number of fermion generations is limited to just a few. In Chapter 39, we discussed how cosmological implications seem to place a limit on the maximum number of neutrino types that can exist, the current best estimate being four. But other theoretical calculations of mass ratios in GUTs suggest that the limit may be only three generations, leaving only one more flavour of quark to be discovered. So, the theoretical minimalists are predicting something of a wilderness once the current generation of excitement is over.

On the other hand, the baroque school of physics can point to the wide range of alternatives to the standard $SU(5)$, including the technicolour theories of dynamical symmetry breaking and the super-symmetric theories of gravity. Both of these could lead to the discovery of new particles and forces in the supposed desert feared by the minima-lists.

Even without adopting these somewhat speculative theoretical alternatives, it is not beyond the bounds of possibility that we should at last reach an energy at which unconfined quarks might be observed, or at which magnetic monopoles might be seen.

But, this is a rather complacent mood in which to leave the subject. As we have said before, history has the knack of repeating itself. In the closing years of the nineteenth-century, no one would have thought that the slight irregularities in the theory of black-body radiation would have led to an entirely new conception of matter as described by the quantum theory. Likewise, few physicists ponder-ing the apparent constancy of the speed of light would have made the connection with $E = mc^2$. Even when quantum theory and relativity were well established, the prediction of antimatter resulting from the synthesis of the two was wholly novel. In the 1930s, few can have realised that the discoveries of the positron and the muon, and the predictions of the pions and the neutrinos were the first heralds of a new order of matter. So, on the basis of history, the only surprise will be if there are no surprises.

41.3 Write-it-yourself

Historical trends may be the only guide to the long-term future, but the next few years are about discoveries and Nobel prizes, and there are lots of them just around the corner. So, I thought the best way to finish this account would be for you, the reader, to write an up-to-date account of these for yourself. Table 41.1 gives you a *proforma* guide to anticipated events which should certainly be reported in the daily press and the scientific maga-zines. Here is a summary of the topics which have been mentioned in the book.

(1) $W^{\pm}$, Z^0 *bosons (Chapter 38)*. These are currently in the process of being discovered at the CERN p$\bar{\text{p}}$ collider. The discovery is a triumph for the experimenters who will not only have confirmed the current confidence in gauge theories, but will have done so with a very elegant and comparatively economical experiment. Other opportunities to

view the bosons will be afforded by the Fermilab Tevatron currently under construction in the USA, operating 1985, and by LEP, operating late 1980s.

(2) *Glueballs (Chapter 31)*. Already, two candidates have been claimed by experimenters at the Brookhaven National Laboratory in New York, USA. In their experiment, they observed the production of two ϕ mesons in collisions. As the ϕ mesons each consist of an s$\bar{\text{s}}$ quark–antiquark pair, and as there are no such quarks in the incoming particles, it means that the new s$\bar{\text{s}}$ pairs must have been created from the vacuum via an intermediate state consisting of gluons. By observing peaks in the mass distribution of the $\phi\phi$ pair, it shows that the gluons form resonances just like the ordinary hadrons. The states are denoted the g resonances at 2.16 and 2.32 GeV mass and have spin 2. If confirmed at other laboratories and if the interpre-tation is accepted, they will become the first fully accredited glueballs, hunted for about a decade since the formulation of QCD.

(3) *Proton decay (Chapter 39)*. Already, can-didates are being offered from the Kolar goldfields and, within the next year or so, other experiments with greater sensitivity will confirm or deny them. If confirmed, proton decay will be enormously signifi-cant as the first verified prediction of grand unifica-tion.

(4) *Higgs particles (Chapters 23 and 38)*. No sign of these has been discovered yet, although they are a firm prediction of the standard electroweak theory. The best immediate chance of spotting them might be as the by-product of Z^0 production at the CERN p$\bar{\text{p}}$ collider or elsewhere in future – or as the decay product of the t$\bar{\text{t}}$ bound state, when it is discovered.

(5) *The top quark (Chapter 36)*. The physics community was somewhat disappointed that the first t$\bar{\text{t}}$ bound state was not discovered in the PETRA accelerator at DESY. It will be very surprising if it lies beyond the top of LEP's maximum planned energy range which may even-tually extend as high as 250 GeV. Before LEP is completed, there is a chance that it will be dis-covered in the mass spectrum of $\mu^+\mu^-$ pairs emerg-ing from hadron–hadron collisions, just as the Υ(b$\bar{\text{b}}$) was discovered.

(6) *Magnetic monopole (Chapter 40)*. Along with gravity waves, this is one of the oldest unveri-

Table 41.1. *The do-it-yourself guide to forthcoming discoveries in particle physics*

Topic	Status (a) Discovered by (b) Date (c) Current status	Crucial measurement	Significance	Nobel rating
$W^\pm$ Z^0 bosons	(a) CERN teams (b) January/May 1983 (c) Candidate events claimed in CERN p$\bar{\text{p}}$ collider, need further samples. Good	$M_W \pm \approx 82$ GeV $M_Z^0 \approx 94$ GeV	Confirms electroweak gauge theory	80%
Glueball	(a) (b) (c) Candidate event claimed by Brookhaven, needs independent confirmation. Good	$M_g \approx 2 \cdot 1$ GeV, spin 2	Supports QCD	40%
Proton decay	(a) (b) (c) Candidates claimed by Indian experiment. Other experiments running but unable to confirm so far. 50–50	$\tau_p \geqslant 6 \times 10^{30}$ years	Dramatic evidence for validity of GUTs	80%
Higgs bosons	(a) (b) (c) Possible at CERN p$\bar{\text{p}}$ 1983–84. Probable at LEP 1987–88	$M_H \leqslant 10$ GeV?	Confirms spontaneous symmetry breaking in gauge theories	40%
Top quark	(a) (b) (c) Possible at higher energy PETRA 1983. Probable at LEP 1987–88	$M_{t\bar{t}} \approx 50$ GeV?	Completes third generation of elementary fermions	30%
Magnetic monopole	(a) (b) (c) One candidate event claimed by Stanford Univ. More needed. Then independent confirmation. Long odds	Existence Mass? Electric charge?	Electric and magnetic charge symmetry predicted by Dirac. Structure of gauge field configurations	90%
Neutrino oscillations and mass	(a) (b) (c) Oscillations perhaps seen in reactor experiment. Mass perhaps detected in nuclear β decay. Long odds	Presence $M_\nu \approx 10$–50 eV?	Structure of GUTs. Eventual fate of the universe	70%
Free quarks	(a) (b) (c) Fractional charges claimed by Stanford Univ. Must be confirmed elsewhere. Quarks or not? Long odds	Existence	Would confuse all current prejudices (ideas)	90%

Table 41.1 (*cont.*)

Topic	Status (a) Discovered by (b) Date (c) Current status	Crucial measurement	Significance	Nobel rating
Super-symmetric or technicolour particles	(a) (b) (c) Possible any time, particularly at PETRA or LEP. Very long odds	Existence	Super-symmetric particles indicate some hope of understanding gravity. Technicolour particles made on eighth day for fun	70%
Gravitational waves (gravitons)	(a) (b) (c) Possible any time. Detectors most probably not yet sensitive enough. Very long odds	Existence	Supports general relativity	90%

fied physics predictions which is still believed. Blas Cabrera's recent claim for a monopole candidate attracted much attention and his experiment at Stanford University in California provides another example of the alternative method to the large-accelerator experiments. But, as it relies on cosmic rays, time alone can establish his case. If other candidates are reported, other similar experiments will set out to confirm them. If they do, then the battle will be on to establish their properites. Their discovery would be a pleasant surprise, possibly confirming some of the more esoteric aspects of the standard electroweak theory.

(7) *Neutrino oscillations and neutrino mass (Chapter 39)*. Currently, the weight of evidence is swinging against the one reactor experiment claiming to have observed the consequences of neutrino oscillations. This tends to confirm the validity of the law of conservation of lepton type. It does not necessarily preclude the existence of a neutrino mass. One nuclear β-decay-type experiment currently claims to have observed a mass for the neutrino of between 14 and 46 eV. Over the next few years other similar experiments will seek to confirm or deny this.

(8) *Detection of free quarks*. One experiment conducted by William Fairbanks at Stanford University in California has made several claims to have observed fractional electronic charges on super-

cooled niobium balls in what is essentially a modern version of Milikan's oil-drop experiments (which first established the value of the electrical charge on the electron). Fairbank's experiment is generally admired and given a fair chance of being correct. But the results suffer from the major barrier-to-credibility problem of why free quarks should exist on Fairbanks' niobium balls and *not* in any of the accelerator experiments. Confirmation of the effect by another experimental team would cause the physics community to look very intensely at the possibility of unconfined quarks. Barring attribution to any unexpected solid-state effects, the free quark would be a major surprise and pose many new theoretical problems.

(9) *Super-symmetric particles, technicolour particles (Chapter 40)*. LEP will doubtless provide the greatest opportunity for the detection of any of the particles predicted by these more speculative theories. A technicolour discovery would indicate the need to revise the Higgs method of mass generation in the standard gauge theories. A super-symmetric discovery would be of enormous significance, indicating both an unexpected relationship between fermions and bosons, quite unexpected from conventional relativistic quantum theory, *and* experimental evidence for the quantum nature of gravity.

(10) *Gravitational waves (Chapter 5)*. A long-

standing prediction of the general theory of relativity; for many years experiments have attempted to detect such waves by monitoring the stresses they would give rise to in ordinary matter as the waves passed through it. The experiments have attempted to detect the gravity waves resulting from extremely violent cosmic happenings (e.g. star collapse to black holes, binary neutron-star rotation, supernovae) by monitoring very sensitive strain gauges attached to large blocks of metal. The effect is broadly similar to the perturbation of a fleet of moored boats when a water wave from a passing craft passes under them.

It is believed currently that the effects of gravity waves are just too feeble to be detected with present experiments. If the waves are detected, then this would be unexpected confirmation of general relativity, and very welcome.

Appendices

Units and constants

Energy: The most common unit of energy in the microworld is the electron volt eV. This is defined as the energy possessed by an electron after it has been accelerated through a potential difference of one volt.

$$1 \text{ eV} = 1.602 \times 10^{-19} \text{ joule}$$

$$1 \text{ MeV} = 10^{6} \text{ eV}$$

$$1 \text{ GeV} = 10^{9} \text{ eV}$$

Mass:

$$\text{electron } m_e = 9.109 \times 10^{-31} \text{ kg}$$
$$= 0.511 \text{ MeV}$$

$$\text{proton } M_p = 1.672 \times 10^{-27} \text{ kg}$$
$$= 939.26 \text{ MeV}$$

Charge:

$$\text{electron } \quad e = 1.602 \times 10^{-19} \text{ coulomb}$$

Speed of light:

$$c = 2.997 \times 10^{8} \text{ m/sec}$$

Planck's constant $\left(\hbar = \dfrac{h}{2\pi} \right)$:

$$\hbar = 1.054 \times 10^{-34} \text{ joule. sec}$$

$$\hbar = 6.582 \times 10^{-22} \text{ MeV. sec}$$

Fine structure constant $\left(\dfrac{e^2}{\hbar c} \right)$:

$$\alpha = \frac{1}{137.036}$$

APPENDIX 2

Glossary

Abelian group: a mathematical group of transformations with the property that the end result of a product of transformations does not depend on the order in which they are performed.

alpha particles: particles first discovered in radioactive α decay and identified as helium nuclei (two protons and two neutrons bound together).

angular momentum: this is the rotational equivalent of ordinary linear momentum, being the product of mass and angular velocity. It is a vector quantity directed along the axis of rotation. In quantum mechanics, angular momentum is quantised in units of Planck's constant (divided by 2π). This corresponds classically to only certain frequencies of rotation being allowed.

antiparticles: particles predicted by combining quantum mechanics and relativity. For each particle there must exist an antiparticle with opposite charge, magnetic moment, and any other internal attributes (lepton, baryon, strangeness, charm quantum numbers), but with identical mass and spin.

asymptotic freedom: the property of the theory of quantum chromodynamics which predicts that the intrinsic strength of the force between quarks decreases as they approach each other.

baryon: a massive, strongly interacting particle with half-integral spin in units of $\hbar$. For instance, the proton, neutron and all their more massive excited resonance states.

beta particles: particles first discovered in radioactive β decay – later identified as electrons.

Big Bang: the theory which asserts that the universe started from a single point of infinite energy density about 10^{10} years ago. Expansion of the universe since the Big Bang is akin to the expansion of the volume within an inflating balloon. There are two main experimental observations supporting the Big Bang hypothesis:

(1) The Hubble expansion: the galaxies always appear to be receding from a point of observation, the more-distant receding faster. The currently observed expansion can be extrapolated backwards in time to date the initial singularity to approximately 10^{10} years ago.
(2) The cosmic microwave background: in 1965 Penzias and Wilson discovered that space is filled with low-frequency photons. The best explanation of which is that they are the doppler-shifted remnants of high-energy photons, released during the Big Bang.

Cabbibo angle: the measure of probability that one flavour of quark (u) will change into other flavours (d, s) under the action of the weak force.

CERN: the Centre of European Nuclear Research, located near Geneva in Switzerland. The resources of European nations are pooled to construct the large and expensive machines needed for high-energy experiments. The major facilities at CERN include the intersecting storage rings (ISR) and the super proton synchrotron (SPS), both of which allow the study of hadron–hadron reactions at very high energies.

charm: the fourth flavour (type) of quark, the discovery of which in 1974 led to a rapid advance in the acceptance of the reality of quarks and in the understanding of their dynamics.

colour: an attribute of quarks which distinguishes otherwise identical quarks of the same flavour.

Three different colours are required: red, green and blue – to distinguish each of the three valence quarks believed to exist within each baryon. It must be stressed that these colours are just labels and have nothing to do with ordinary colour. Colour is now realised to be the source of the interquark forces, and so the three colours can be thought of as three different colour charges analogous to electric charge.

cross-section: this is the basic measurement of the probability of particles interacting. It is expressed as an effective target area (for instance, cm^2), which can be related to the quantum-mechanical probability of interacting by multiplying by factors such as the flux of particles entering the interaction region. The basic unit for nuclear physics is called the barn (b) 1 b $= 10^{-24}$ cm^2. Typical hadron collisions are measured in millibarns (mb) 1 mb $= 10^{-27}$ cm^2. Typical neutrino collision cross-sections are about 10^{-39} cm^2.

DESY: the German national high-energy laboratory near Hamburg. Home of DORIS and PETRA which are both electron–positron storage rings and both of which have enjoyed successful experiments over the last decade.

deuteron: the nucleus of deuterium, an isotope of hydrogen. It consists of one proton and one neutron bound together.

diffraction: a property which distinguishes wave-like motions. When a wave is incident upon a barrier which is broken by a small slit, and if the dimension of the slit is comparable to the wavelength, then the slit will act as a new, isotropic source of the wave motion. This is to be contrasted with the behaviour of particle or ray motions in which the presence of the slit will collimate the motion into a single direction.

dimensionality: physically significant quantities have some dimensions associated with them. These dimensions are those of mass M, length L, and time T and any combinations of their powers. So, for instance, momentum has dimensions of mass $\times$ velocity (MLT^{-1}) and energy has the dimension of force $\times$ distance ($ML^2 T^{-2}$). Equations describing physical quantities can be checked by ensuring they have the correct dimensions. Sometimes, it is possible to define quantities in which the dimensions cancel out. These dimensionless quantities are significant as they are entirely independent of the conventions used to define mass, length and time.

eigenstate, eigenvalue: the eigenvalue of a matrix **M** is a number λ which satisfies the equation

$$\mathbf{M}\psi = \lambda\psi,$$

where ψ is called an eigenvector of the matrix. In quantum mechanics, the matrix **M** will represent an operator (i.e. some act of 'measurement': of position, momentum, energy) and λ will correspond to the value of that particular variable if the wavefunction is in an eigenstate of the operator (i.e. if the above equation is satisfied).

elastic scattering: particle reactions in which the same particles emerge from the reaction as entered it (e.g. π^-p $\rightarrow \pi^-$p). Because of this, conventional three-momenta and energy are conserved separately. In inelastic scattering, in which different or new particles emerge, this is no longer true as energy has been used to alter or create new particles.

fermions: particles with half-integral spin in units of $\hbar$ and which obey Pauli's exclusion principle.

flavours: the name given to the quark types up, down, strange, charm, top and bottom. These govern the weak interaction dynamics of the quarks, the flavours being essentially the 'weak charges' on the quarks.

Fourier sum: the sum of a collection of wave motions. At some points, peaks will add together to form larger peaks and at other points, peaks will cancel with troughs to form zeros (or nodes) in the wave amplitude. In general, the result will be an arbitary form of amplitude distributed throughout space.

gamma-ray: rays first discovered in radioactive material, later identified as high-energy photons.

gauge theory: the term used to describe those theories whose dynamics originate from the invariance of the formulae (especially the Lagrangian) under symmetry or 'gauge' transformations. For instance, in electrodynamics the theory is invariant under redefinition of the electrostatic potential, which corresponds eventually to the law of conservation of electric charge. In quantum electrodynamics, this is rephrased as invariance under transformations redefining the phase of the electron waves. The name gauge theory is archaic, deriving from earlier theories based on invariance under transformations of scale of measurement or gauge.

gluons, glueballs: gluons are the gauge particles of quantum chromodynamics which carry the strong colour force from one quark to another. Because of the structure of the theory, gluons can also interact with themselves and form particles consisting only of gluons bound together. These particles, which have not yet been finally confirmed, are called glueballs.

Goldstone bosons: particles named after the Cambridge physicist Jeffrey Goldstone, which must exist as a consequence of the spontaneous breaking of a global gauge symmetry.

graviton: the hypothetical quantum of the gravitational field. It is the gravitational relation of the photon, the gluons and the $W^{\pm}$, Z^0 bosons. Because of the extremely feeble nature of gravity, it is difficult to foresee evidence for the existence of the graviton being obtained.

group theory: this is the branch of mathematics which describes symmetry. A mathematical group **G** is defined as a collection of elements $(a, b, c \ldots)$ with the properties:

(1) If a and b are in group **G**, then the product of the two elements, ab, is also in **G**.
(2) There exists a unit element e such that $ae = a$ for all elements in **G**.
(3) Each element a has an inverse a^{-1} such that $aa^{-1} = e$.

So, for instance, the rotations of an (xy) coordinate system about the z axis form a group as the effect of any two rotations $\theta_1 \theta_2$ is equivalent to the effect of one big rotation θ_3. Such a group is referred to as a continuous group as the parameters of the elements vary continuously over a range of values.

There are other so-called discrete groups in which a finite number of individual elements satisfy the group conditions (1) and (3).

In general, the elements (or transformations) of any group may be written as matrices connecting the old basis of states on which the group acts (i.e. the coordinate system) with the new transformed basis. These matrices are referred to as representations of the group.

When a physical system is found to be governed by a symmetry group, i.e. when its equations of motion are left invariant under transformations of the group, then the representations of the group specify the multiplicities existing within the system. (The order of the matrices of the irreducible representations corresponds to the degeneracies of the eigenvalues of the operators governed by the symmetry.)

In the case of the continous dynamical symmetries of the Lagrangians of particle theories these multiplicities correspond to the number of gauge bosons needed to effect the transformations between the basis states (the matter fields), i.e. to carry the forces between them.

In the case of the discrete internal symmetries of hadrons, the multiplicities give the number of particles and their quantum numbers for each value of spin and parity assignments at the same mass (assuming perfect symmetry).

hadron: any particle which experiences the strong nuclear force.

helicity: the projection of a particle's spin along its direction of motion. The helicity of a particle is described as being either left- or right-handed depending on whether its spin vector is in the direction of motion or against it. See Figure 13.3.

Higgs mechanism: the method in the gauge theory of the weak force which gives mass to the $W^{\pm}$ and Z^0 bosons. New 'Higgs' fields are introduced to the theory in an originally locally gauge invariant way. Redefinition of the Higgs fields then breaks the local gauge symmetry, generates the mass of the $W^{\pm}$ and Z^0 bosons and gives rise to an observable remnant Higgs particle.

hyperon: a baryon with a non-zero strangeness quantum number.

isotopic spin or isospin: this concept describes the electric charge of a particle by the orientation of an hypothetical spin vector in an abstract charge space. The charge independence of nuclear forces is then described by the invariance of the equations describing them under rotations of the isospin vector in this abstract charge space.

K meson or kaon: the name of the spin-0 mesons with non-zero strangeness quantum numbers.

Lagrangian: a mathematical expression describing the energy of a physical system. It is possible to derive the dynamical equations of motion of the systems from the Lagrangian by a mathematical procedure known as Hamilton's variational principle.

lepton: a particle which does not experience the strong nuclear force. Taken to refer specifically to the electron, the muon and the tau leptons and their respective neutrinos (making six leptons in total), and their antiparticles. The name was coined originally to refer to a light particle.

lifetime: if the probability of a particle decaying per unit time is λ, then the number decaying dN in an infinitesimal interval dt is given by

$$dN = \lambda N dt.$$

If a collection of N_0 particles is present at $t=0$, then the number surviving to time t will be

$$N = N_0 e^{-\lambda t}.$$

The lifetime τ of the particles is defined as the reciprocal of the decay probability per unit time

$$\tau = \frac{1}{\lambda},$$

which is the time by which a sample of particles will have decreased to $1/e$ of its initial population. It is also the mean lifetime of the sample. It is related to the half-life time of the sample $\tau_{\frac{1}{2}}$ by:

$$\tau_{\frac{1}{2}} = \ln 2 . \tau.$$

magnetic moment: this is a measure of the extent to which a physical system (e.g. an atom, nucleus, particle) looks like a north–south magnetic pole pair. It is generally measured in units of magnetons $((e\hbar)/(2\,mc))$. When the electron mass is used, this is known as the Bohr magneton. When the proton mass is used, it is known as the nuclear magneton.

magnetic monopole: a hypothetical object proposed by Dirac which consists of only one magnetic pole, in contrast to all known sources of magnetism which consist of north–south pole pairs or combinations of them. It is currently believed that if such objects do exist they must be very massive ($O[100\ M_p]$) to explain why they have not been observed in high-energy experiments to date.

mass-shell: in quantum mechanics, a particle's energy and its momentum are essentially independent of each other. A particle is said to be 'on mass-shell' when its energy and momentum satisfy the formula from special relativity

$$E^2 = p^2\,c^2 + m_0^2\,c^4,$$

which is necessary for it to exist as an observable particle.

meson: a hadron with integer spin in units of $\hbar$, such as the pion and the kaon. The name was coined originally to refer to the middleweight particles between the light lepton and the heavy baryons.

muon: a second-generation lepton. To all intents and purposes, it is simply a more massive brother of the electron.

natural units: in formulae describing physical quantities, it is possible to set h, Planck's constant, and c, the speed of light, equal to unity (equivalent to dividing by appropriate powers of h and c). This gives the quantity in dimensionless or 'natural' units which make comparisons between quantities clearer and allows less-cluttered theoretical computations. Of course, when theoretical calculations are to be compared with experimental measurements, the correct values of h and c must be taken into account.

neutral current reactions: weak interaction reactions in which no electric charge is exchanged between the colliding particles. Observation of such reactions in

1973 provided important support for the then-developing gauge theory of the weak interactions.

neutron: one of the constituents of the atomic nucleus discovered in 1932. Because it is electrically neutral, it must be bound into the atomic nucleus by an unfamiliar strong nuclear force. Also, the neutron's slow decay is indicative of yet another weak nuclear force. Despite being electrically neutral, the neutron does possess both an electric dipole moment (as if it were made of a positive and a negative electric charge separated by a minute distance) and a magnetic moment, both of which are indicative of some internal electrical structure.

neutrino: an electrically neutral, massless particle of spin $\frac{1}{2}$, which interacts only by the weak interaction and gravity. It was first postulated by Pauli in 1930 to ensure conservation of energy and of angular momentum in nuclear β decay. Now, three separate types of neutrino are known to exist to match the three types of massive lepton (the electron, the muon and the tau lepton).

Noether's theorem: a mathematical theorem which states that for every symmetry of the Lagrangian of a physical system (i.e. for every set of transformations redefining the quantities in the Lagrangian under which it is invariant) there will be some quantity which is conserved by the dynamics of the system.

nucleon: a generic name for the proton and the neutron.

parity: the quantum-mechanical operation which reverses the signs of the coordinate axes being used to describe a system. It is equivalent to a mirror reflection followed by a rotation through 180°.

parton: a generic term used to describe any particle inside the nucleons. Both quarks and gluons are included.

photon: the quantum of the electromagnetic field. It has spin 1, in units of $\hbar$ and is massless. Virtual photons which carry the electromagnetic force between charged particles can adopt an effective mass for a short period in accordance with Heisenberg's uncertainty principle.

positron: the antiparticle of the electron discovered by Anderson in 1934. It has the same mass and spin as the electron, but opposite charge and magnetic moment.

propagator: the mathematical expression used to describe the propagation in space–time of virtual particles. It fulfils the role for virtual particles as does the wavefunction for real particles.

proton: one of the constituents of the atomic nucleus which carries positive quantum of electronic charge and spin $\hbar/2$. It is the least-massive baryon and, as a result, is the particle into which all other baryons will eventually decay. Until recently it was believed to be absolutely stable, but the most modern theories predict that it will decay very, very slowly. Such decays are currently the subject of intense experimental searches.

quantum chromodynamics (QCD): the quantum field theory describing the interaction of quarks through the strong 'colour' field (whose quanta are gluons).

quantum electrodynamics (QED): the quantum field theory describing the interactions of electrically charged particles through the electromagnetic field.

quantum field theory: the theory used to describe the creation and destruction of particles in elementary particle reactions. The name arises because the theory introduces the idea that quantum fields are the ultimate reality and that particles are merely the localised quanta of these fields, which can be created, destroyed or transformed as easily as can knots in a rope.

quantum flavourdynamics (QFD): the alternative name for the Glashow–Weinberg–Salam theory (also a quantum field theory) describing the weak interactions of both hadrons and leptons via the $W^{\pm}$ and Z^0 bosons; so-called because the weak interaction alters the flavours of quarks participating, according to the Cabbibo rule.

quantum theory: the theory used to describe anything very small, typically of atomic dimensions or less. The name arises from the principal distinguishing feature of the theory, which is that its description of energy does not allow continuously divisible amounts, but requires there to exist minimum amounts, called quanta, which cannot be divided.

Regge theory: a description of the strong interactions of hadrons used before (and currently as a macroscopic approximation to) the QCD description of underlying quark dynamics. The theory describes the strong force by the exchange of a regge trajectory rather than a single meson. A reggeon trajectory or reggeon, is the series of mesons of increasing mass and increasing spins but with otherwise identical quantum numbers.

renormalisation: the process in quantum field theory required to ensure that the basic quantities being described (e.g. in QED: the electron, the photon, and the quantum of electric charge) are sufficiently accurately defined to avoid mathematical irregularities (such as infinites) arising in physical predictions.

resonance particles or resonances: hadronic particles which exist for only a very brief time (10^{-23} seconds) before decaying into stable hadrons.

r f power: electric and magnetic fields alternating at the frequencies of radio waves (up to 10^{10} Hz) which can be used to accelerate charged particles in accelerators.

scaling: the phenomenon observed in deep inelastic scattering, first predicted by James Bjørken, whereby the structure functions which describe the shape of the nucleon depend not on the energy or momentum involved in the reaction, but on some dimensionless ratio of the two. The structure functions become independent of any dimensional scale.

second quantisation: the mathematical procedure in quantum field theory which represents material particles as localised excitations (or quanta) of the underlying fields on which the theory is based.

SLAC: the acronym for the Stanford Linear Accel-

erator Center at Stanford University in California. It is distinguished by having a two-mile-long accelerator in which electrons and positrons can be accelerated for subsequent injection into storage rings in which their interactions can be observed. It was in the SPEAR rings at SLAC in which the ψ (psi) meson and the τ (tau) lepton were first observed in the mid-1970s, which heralded the 'new physics' – a grand name for our increased confidence in the validity of the quark–lepton generation structure of the microworld. The newer, PEP, rings at SLAC are currently in operation.

spin: many particles possess an intrinsic angular momentum which can be thought of as resulting from the particles spinning about an axis passing through their centres. This spin angular momentum is observed to be quantised in integral or half-integral units of $\hbar$ (and predicted to be so for spin $-\frac{1}{2}$ particles by the Dirac equation), which, viewed classically, corresponds to only certain frequencies of rotation being allowed. Fundamentally, spin describes how the particle fields transform under the transformations of special relativity.

spontaneous symmetry breaking: any situation in physics in which the state of minimum energy of a physical system is not symmetric under certain transformations of the coordinates of the system. In evolving towards the state of minimum energy, symmetry will be lost. So, for instance, the state of minimum energy for an iron magnet is that in which all the atomic spins are aligned in the same direction, giving rise to a net macroscopic magnetism. But, if the energy of the system is raised, the symmetry may be restored (the application of heat to an iron magnet will destroy the asymmetric macroscopic magnetic field).

strangeness: a quantum number assigned to the elementary particles which are conserved by the strong nuclear force. Originally, strangeness was introduced to explain why some hadrons decay quickly by the strong nuclear force whilst others decay slowly by the weak force. Strangeness is currently recognised to be the third of the quark flavours.

tensor: a quantity whose components may be

described by an arbitrary number of indices A_{ijk} .... In this context, an ordinary vector V_i ($i = 1$, 2, 3 in 3-dimensional space) can be described as a tensor of the first rank and a 2-dimensional matrix M_{ij} as a tensor of the second rank.

top: the sixth, as yet undiscovered, flavour of quark.

van der Waals forces: these are the very weak electromagnetic forces existing between electrically neutral atoms. They result because the individual electric charges within any atom may have greater or lesser effects on the charges within a nearby atom, due to the spatial distribution of the charges. The forces decline very rapidly with increasing separation of the atoms ($\alpha(1/r^7)$).

virtual processes: processes which do not conserve energy and momentum over microscopic timescales as allowed by Heisenberg's uncertainty principle. These processes cannot be observed but form the transient intermediate states of elementary particle processes which are observed and which do conserve energy and momentum.

wavefunction: a particle in quantum mechanics is completely described by its wavefunction, denoted ψ. This is a mathematical expression which describes the propogation in time and space of the particle by a factor such as $e^{i(kx - wt)}$, and describes properties such as spin, isospin, flavour and colour by factors giving the orientation of vectors in the appropriate real and abstract spaces involved. The wavefunction squared gives the probability of finding the particle at a particular point with particular values of its other attributes.

APPENDIX 3

List of symbols

a acceleration

$A_1 A_2$ baryon resonances

b bottom quark flavour
     blue quark colour

B baryon number
     gauge field of $SU(2)^W$

c speed of light
     charm quark flavour

C charge conjugation operator

d down quark flavour

D

e electron
     electronic charge

E energy

f parton probability distribution

$F_1 \ F_2 \ F_3$ deep inelastic structure functions

F meson
     force

g green quark colour
     g factor of the electron
     general coupling constant

$\mathbf{G}$ Group

G Newton's constant

G_F Fermi's coupling constant

h Planck's constant
hadrons
strangeness-conserving weak hadronic current

H magnetic field

H^W weak hadronic current

I isospin

I_3 third component of isospin

J meson

J^W total weak current

K constant of proportionality
strange meson

$l_e\ l_\mu$ electronic or muonic lepton

L^W weak leptonic current

L Lagrangian

$\mathscr{L}$ Lagrangian density

m mass
quantum-mechanical amplitude of sub-processes

M total quantum-mechanical amplitude of a process

$\mathbf{M}$ matrix

n neutron

N numbers of . . .

$N_L\ N_R$ number of left-spinning, (right-spinning)

N^* baryon resonances

p magnitude of momentum
proton

$\mathbf{p}$ momentum vector

P probability (of occurrence of quantum-mechanical event)

$\mathbf{P}$ parity operator

P polarisation
fraction of nucleon momentum carried by the quark

q quark

q^2 momentum transfer squared in deep inelastic scattering

Q electric charge (in units of e)

r magnitude of distance
red quark colour

R vacuum expectation value of Higgs fields
ratio of cross-sections for e^+e^- into hadrons to e^+e^- into muon–antimuon pair

s space–time interval
strange quark flavour

s_3 third component of spin

$s^\pm$ strangeness-changing weak hadronic current

S strangeness quantum number

$SU(2)$ special unitary groups of transforma-
$SU(3)$ tions of order 2, 3 and 5 respectively
$SU(5)$

t time
top quark flavour

$\mathbf{T}$ time-reversal operator

u magnitude of velocity
up quark flavour

v magnitude of velocity

$\mathbf{v}$ velocity vector

V symbol for V particle (archaic)

$W^\pm$ W boson

$\mathbf{x}$ position vector

x deep inelastic scattering variable
spatial separation

X massive gauge bosons predicted by GUTs
unspecified final state of reaction

y fraction of energy transferred in deep inelastic scattering

Y hypercharge quantum number

Z^0 Z boson

α (alpha) α radiation or particles (helium nuclei)
electromagnetic fine structure constant

β (beta) β radiation or particles (electrons)

γ (gamma) γ radiation or particles (photons)

Γ Fermi's weak interaction couplings

δ (delta) infinitesimal increment in variable in calculus

Δ infinitesimal amount of ...
delta baryon

ε (espilon) small number

ζ (zeta)

η (eta)

θ (theta) theta meson (archaic)
an angle
the Cabbibo angle
the weak angle

ι (iota)

κ (kappa)

λ (lambda) wavelength

Λ lambda hyperon

μ (mu) muon

ν (nu) neutrino
deep inelastic scattering energy transfer
frequency

ζ (xi)

Ξ Xi hyperon

o   (omicron)

π (pi) 3.141 5927
pion

ρ (rho) rho meson
hypothetical hadronic isospin gauge particle

σ (sigma) cross-section

$\sum$ summation over ...

Σ sigma hyperon

τ (tau) tau heavy lepton
tau meson (archaic)
lifetime, duration

υ (upsilon)

Υ upsilon meson

ϕ (phi) Higgs particles

χ (chi) χ meson

ψ (psi) quantum-mechanical wavefunction
psi meson

ω (omega) omega meson

Ω^- omega minus baryon

$=$ is equal to

$\equiv$ is identical with

$\approx$ is approximately equal to

$a > b$ a is greater than b

$a < b$ a is less than b

$a \leqslant b$ a is equal to, or less than, b

$a \geqslant b$ a is equal to, or greater than, b

$a \supset b$ a is included in b

$\langle i \,|\, M \,|\, f \rangle$ initial i and final f states connected by a quantum-mechanical amplitude

$y \bigcirc [X]$ y is of the same order as X

APPENDIX 4

Bibliography

In this brief guide, I have provided two sets of references on most of the subjects concerned. The first, non-specialist, category includes articles and books accessible to the audience of books such as this one. The specialist category includes material for the professional student of physics and is generally aimed at the level of a third-year undergraduate or first-year post-graduate.

(NS) denotes the non-specialist category.

(S) denotes the specialist category.

Part 0

(NS) *From X-rays to Quarks*, Emilio Segrè. W.H. Freeman, San Franciso, 1980.

(NS) *Relativity*, Albert Einstein. Methuen, London, 1920.

(S) *Special Relativity*, A.P. French. Thomas Nelson, 1968.

(S) *Simple Quantum Physics*, Peter Landshoff & Allen Metherell. Cambridge University Press, 1979.

(S) *Relativistic Quantum Mechanics*, I.J.R. Aitchison. Macmillan, London, 1972.

(S) *Relativistic Quantum Mechanics and Relativistic Quantum Fields* (two volumes), James D. Bjørken & Sidney Drell. McGraw-Hill, New York, 1964.

Part 1 and general

(NS) *The Forces of Nature*, P.C.W. Davies. Cambridge University Press, 1979.

(NS) *The Particle Play*, J.C. Polkinghorne. W.H. Freeman, Oxford, 1979.

(NS) *The Nature of Matter*, J.H. Mulvey (ed.). Clarendon Press, Oxford, 1981.

(S) *The Physics of Elementary Particles*, L.J. Tassie. Longman, London, 1973. (A good introductory textbook.)

(S) *An Introduction to High-Energy Physics*, Donald H. Perkins. Addison Wesley (2nd edn), 1982. (An excellent textbook on most of the material mentioned in this book.)

(S) *Symmetry Principles in Elementary Particle Physics*, W.M. Gibson & B.R. Pollard. Cambridge University Press, 1976.

Part 2

(NS) 'Resonance particles', R.D. Hill. *Scientific American*, **208** (1), January 1963.

(NS) 'Strongly interacting particles', Geoffery F. Chew, Murray Gell-Mann & Arthur H. Rosenfeld. *Scientific American*, **210** (2), February 1964.

(NS) 'Dual-resonance models of elementary particles', John H. Schwarz. *Scientific American*, **232** (5), 61–7, February 1975.

(S) *High Energy Hadron Physics*, Martin L. Perl. John Wiley, New York, 1974. (A thorough textbook on pre-QCD hadron physics.)

(S) 'High energy hadron collisions: a point of view', J.P. Aurenche & J.E. Paton. *Reports on Progress in Physics*, **39** (2), February 1976.

Parts 3 and 4

(NS) 'The weak interactions', S.B. Treiman. *Scientific American*, March 1959.

(NS) 'The two-neutrino experiment', Leon M. Lederman. *Scientific American*, **208** (3), March 1963.

(NS) 'Violations of symmetry in physics', Eugene P. Wigner. *Scientific American*, **213** (6), 28–36, December 1965.

(NS) 'Weak interactions', Mary K. Gaillard. *Nature*, **279**, 585–9, June 1979.

(S) *Weak Interactions*, D. Bailin. Adam Hilger, Bristol (2nd edn), 1982.

Part 5

(NS) 'Unified theories of elementary-particle

interaction', Steven Weinberg. *Scientific American*, **231** (1), 50–9, July 1974.

(NS) 'Gauge invariance in physics', G.'tHooft. *Scientific American*, **242** (6), 90–116.

(S) *Gauge Theories in Particle Physics*, I.J.R. Aitchison & A.J.G. Hey. Adam Hilger, Bristol, 1982.

(S) *Gauge Theories of Weak Interactions*, J.C. Taylor. Cambridge University Press, 1978.

Part 6

(S) 'Inelastic lepton-nucleon scattering', D.H. Perkins. *Reports on Progress in Physics*, **40**, 409–81, 1977.

(S) *An Introduction to Quarks and Partons*, F.E. Close. Academic Press, London, 1979.

Part 7

(NS) 'Quantum chromodynamics', William Marciano & Heinz Pagels. *Nature*, **279**, 479–83, June 1979.
and

(S) *Physics Reports*, **36C**, 137, 1978.

(NS) 'Quark confinement', R.L. Jaffe. *Nature*, **268**, 201–9, July 1977.

(S) *Lectures on Lepton Nuclear Scattering and Quantum Chromodynamics*, W.B. Atwood, J.D. Bjørken, S.J. Brodsky & R. Stroynowski. Birkhauser, Boston, 1982.

Part 8

(NS) 'Electron–positron collisions', Alan M. Litke & Richard Wilson. *Scientific American*, **229** (4), 104–13, October 1973.

(NS) 'Fundamental particles with charm', Roy F. Schwitters. *Scientific American*, **237** (4), 56–70, October 1977.

(NS) 'The upsilon particle', Leon M. Lederman. *Scientific American*, **239** (4), 60–8, October 1978.

(NS) 'The tau heavy lepton', Martin L. Perl. *Nature*, **275**, 273–7, September 1978.

Part 9

(NS) 'The search for intermediate vector bosons', David B. Cline, Carlo Rubbia & Simon van der Meer. *Scientific American*, 1982.

(S) 'Antiproton–proton colliders and intermediate bosons', David B. Cline &

Carlo Rubbia. *Physics Today*. August 1980.

(NS) 'A unified theory of elementary particles and forces', Howard Georgi, *Scientific American*, **244** (4), 40–55, April 1981.

(NS) 'Unified theory of elementary-particle forces', Howard Georgi & Sheldon L. Glashow. *Physics Today*, September 1980.

(NS) 'The cosmic asymmetry between matter and antimatter', Frank Wilczek. *Scientific American*, **243** (6), 60–8, December 1980.

(NS) 'Cosmology and elementary particle physics', Michael S. Turner & David N. Schramm. *Physics Today*, September 1979.

(NS) 'Supergravity and the unification of the laws of physics', Daniel Z. Freedman & Peter van Nieuwenhuizen. *Scientific American*, **238** (2), 126–43, February 1978.

(S) *Supergravity '81*, S. Ferrara & J.G. Taylor (eds.). Cambridge University Press, 1982.

(NS) 'Solitons', Claudio Rebbi, *Scientific American*, **240** (2), 76–92, February 1979.

(S) *Monopoles in Quantum Field Theory*, N.S. Craigie, P. Goddard, W. Nahm (eds.). World Scientific Publishing Co., Singapore.

(S) 'Third workshop on grand unification', Paul H. Frampton, Sheldon L. Glashow & Hendrik van Dam (eds.). Birkhauser, Boston 1982.

General sources of information

For the non-specialist interested in keeping abreast of the modern developments in particle physics, the following magazines and journals combine to form a good coverage: *New Scientist, Scientific American, Nature, Physics Today* and *CERN Courier*.

For the specialist, the most up-to-date letters appear in *Physical Review Letters* and *Physics Letters*. The mainstream journals on the subject are *The Physical Review (D)*, *Nuclear Physics (B)* and, to a lesser extent, *Nuovo Cimento* and *Journal of Physics (G)*. Review articles on the subject generally appear in *Physics Reports (C)*, *Reviews of Modern Physics* and *Reports on Progress in Physics*. For up-to-date, comprehensive reviews of the entire field the best sources are the summer school lecture note compilations and the conference proceedings on the subject published periodically; generally by CERN or SLAC.

APPENDIX 5

Abridged particle table

The following list is a complete account of the particles known to date. Only the main features of the particle decays have been listed. No long-standing unconfirmed particles have been included, nor have the various dibaryon states identified to date. For fuller details see the *Review of Particle Properties*, published by the Particle Data Group at CERN (*Physics Letters*, **111B**, April 1982 and reprinted by CERN).

Table A.5. *Abridged particle table*

Particle	Parity P Isospin I Spin J $I\,(J^P)$	Mass (MeV)	Lifetime τ (seconds) Width (MeV)Γ	Decays Main mode	Fraction %
INTERMEDIATE BOSONS					
γ	$0(1^-)$	$0\ (<6\times10^{-22})$	Stable		
$W^\pm$	$1(1^-)$	$81\,000\pm5000$	?	Tentative discovery 1983	
Z^0	$1(1^-)$	$\sim94\,000$	?	Tentative discovery 1983	
LEPTONS					
ν_e	$J=\tfrac{1}{2}$	$0\ (<0.000\,046)$	Stable		
e	$J=\tfrac{1}{2}$	$0.511\,003\,4\pm0.000\,001\,4$	Stable		
ν_μ	$J=\tfrac{1}{2}$	$0\ (<0.52)$	Stable		
μ	$J=\tfrac{1}{2}$	$105.659\,43\pm0.000\,18$	$\tau=2.197\,14\times10^{-6}$ $\pm0.000\,07$	$\bar\mu\to e^-\bar\nu\nu$ $e^-\bar\nu\nu\gamma$	98.6 1.4
ν_τ	$J=\tfrac{1}{2}$	<250			
τ	$J=\tfrac{1}{2}$	1784.2 ± 3.2	$\tau=(4.6\pm1.9)\times10^{-13}$	$\tau^-\to\mu^-\bar\nu\nu$ $\to e^-\bar\nu\nu$ $\to$ hadrons	19 17 64
NON-STRANGE MESONS					
$\pi^\pm$	$1(0^-)$	139.5673 ± 0.0007	$\tau=2.6030\times10^{-8}\pm0.0023$	$\pi^+\to\mu^+\nu$	99.99

Table A.5. *Abridged particle table*

Particle	Parity P Isospin I Spin J $I(J^P)$	Mass (MeV)	Lifetime τ (seconds) Width (MeV)Γ	Decays Main mode	Fraction %
π^0	$1(0^-)$	134.9630 ± 0.0038	$\tau = (0.83 \pm 0.06) \times 10^{-16}$	$\gamma\gamma$ $\gamma e^+ e^-$	98.8 1.2
η	$0(0^-)$	548.4 ± 0.6	$\Gamma = 0.83$ keV ± 0.12 keV	$\gamma\gamma$ $3\pi^0$ $\pi^+\pi^-\pi^0$	39.1 31.8 23.7
$\rho(770)$	$1(1^-)$	769 ± 3	$\Gamma = 154 \pm 5$	$\pi\pi$	≈ 100
$\omega(783)$	$0(1^-)$	782.6 ± 0.2	$\Gamma = 9.9 \pm 0.3$	$\pi^+\pi^-\pi^0$ $\pi^0\gamma$	89.9 8.7
$\eta'(958)$	$0(0^-)$	957.57 ± 0.25	$\Gamma = 0.28 \pm 0.10$	$\eta\pi\pi$ $\rho^0\gamma$	65.3 30.0
s*(975)	$0(0^+)$	975 ± 4	$\Gamma = 33 \pm 6$	$\pi\pi$ $K\bar{K}$	78 22
$\delta(980)$	$1(0^+)$	983 ± 2	$\Gamma = 54 \pm 7$	$\eta\pi$ $K\bar{K}$	? ?
$\phi(1020)$	$0(1^-)$	1019.61 ± 0.07	$\Gamma = 4.21 \pm 0.13$	K^+K^- $K_{LONG}K_{SHORT}$ $\pi^+\pi^-\pi^0$	49.1 34.6 14.8
H(1190)	$0(1^+)$	1190 ± 60	$\Gamma = 320 \pm 50$	$\rho\pi$	?
B(1235)	$1(1^+)$	1233 ± 10	$\Gamma = 137 \pm 10$	$\omega\pi$	?
f(1270)	$0(2^+)$	1273 ± 5	$\Gamma = 179 \pm 20$	$\pi\pi$ $2\pi^+2\pi^-$ $K\bar{K}$	83.1 2.8 2.9
$A_1(1270)$	$1(1^+)$	1275 ± 30	$\Gamma = 315 \pm 45$	$\rho\pi$	most
D(1285)	$0(1^+)$	1283 ± 5	$\Gamma = 26 \pm 5$	$K\bar{K}\pi$ $\eta\pi\pi$ $\delta\pi$	11 49 36
e(1300)	$0(0^+)$	≈ 1300	$\Gamma = 200–600$	$\pi\pi$	most
$\pi(1300)$	$1(0^-)$	1300 ± 100	$\Gamma = 200–600$	Tentative assignment	
$A_2(1320)$	$1(2^+)$	1318 ± 5	$\Gamma = 110 \pm 5$	$\rho\pi$ $\eta\pi$ $\omega\pi\pi$	70.1 14.5 10.6
E(1420)	$0(1^+)$	1418 ± 10	$\Gamma = 52 \pm 10$	$K\bar{K}\pi$	most
f'(1515)	$0(2^+)$	1520 ± 10	$\Gamma = 75 \pm 10$	$K\bar{K}$	most

Table A.5. *Abridged particle table*

Particle	Parity P Isospin I Spin J $I(J^P)$	Mass (MeV)	Lifetime τ (seconds) Width (MeV)Γ	Decays Main mode	Fraction %
$\rho'(1600)$	$1(1^-)$	1600 ± 20	$\Gamma = 300 \pm 100$	4π $\pi\pi$	most < 30
$\omega(1670)$	$0(3^-)$	1688 ± 5	$\Gamma = 166 \pm 15$	$3\pi, 5\pi$	most
$A_3(1680)$	$1(2^-)$	1680 ± 30	$\Gamma = 250 \pm 50$	$f\pi$ $\rho\pi$	55 36
$\phi'(1680)$	$0(1^-)$	1684 ± 15	$\Gamma = 126 \pm 22$	$K^*\bar{K} + \overline{K^*}K$	most
$g(1690)$	$1(3^-)$	1691 ± 5	$\Gamma = 200 \pm 20$	$2\pi, 4\pi$	most
$h(2040)$	$0(4^+)$	2040 ± 20	$\Gamma = 150 \pm 50$	$\pi\pi$ $K\bar{K}$	? ?
$\eta_c(2980)$	$0(\quad)$	2981 ± 6	$\Gamma < 20$	$\eta\pi^+\pi^-$ others	?
$J/\psi(3100)$	$0(1^-)$	3096.9 ± 0.1	0.063 ± 0.009	e^+e^- $\mu^+\mu^-$ hadrons	7.4 7.4 85
$\chi(3415)$	$0(0^+)$	3415.0 ± 1.0	?	$2(\pi^+\pi^-)$ $\pi^+\pi^-K^+K^-$	4.3 3.4
$\chi(3510)$	$0(1^+)$	3510.0 ± 0.6	?	$\gamma\psi$ $3(\pi^+\pi^-)$	28 2.4
$\chi(3555)$	$0(2^+)$	3555.8 ± 0.6	?	$\gamma\psi$ $2(\pi^+\pi^-)$	15.7 2.3
$\psi(3685)$	$0(1^-)$	3686.0 ± 0.1	$\Gamma = 0.215 \pm 0.040$	e^+e^- $\mu^+\mu^-$ hadrons	0.9 0.8 98.1
$\psi(3770)$	$?(1^-)$	3770 ± 3	$\Gamma = 25 \pm 3$	$D\bar{D}$	most
$\psi(4030)$	$?(1^-)$	4030 ± 5	$\Gamma = 52 \pm 10$	hadrons	most
$\psi(4160)$	$?(1^-)$	4159 ± 20	$\Gamma = 78 \pm 20$	hadrons	most
$\psi(4415)$	$?(1^-)$	4415 ± 6	$\Gamma = 43 \pm 20$	hadrons	most
$\Upsilon(9460)$	$?(1^-)$	9456 ± 10	$\Gamma = 0.042 \pm 0.015$	$\mu^+\mu^-$ e^+e^-	3.2 2.8
$\Upsilon(10\,020)$	$?(1^-)$	$10\,016 \pm 10$	$\Gamma = 0.030 \pm 0.010$	$\Upsilon\pi\pi$ e^+e^- $\mu^+\mu^-$	30 1.7 seen

Table A.5. *Abridged particle table*

Particle	Parity P Isospin I Spin J $I(J^P)$	Mass (MeV)	Lifetime τ (seconds) Width (MeV)Γ	Decays Main mode	Fraction %
$\Upsilon(10\,350)$	$?(1^-)$	$10\,347\pm10$	?	e^+e^-	seen
$\Upsilon(10\,570)$	$?(1^-)$	$10\,569\pm10$	$\Gamma=14\pm5$	e^+e^-	seen
STRANGE MESONS					
$K^\pm$	$\frac{1}{2}(0^-)$	493.667 ± 0.015	$\tau=(1.2371\pm0.0026)\times10^{-8}$	$K^+\to\mu^+\nu$ $\pi^+\pi^0$	63.5 21.2
K^0 $\bar{K}^0$	$\frac{1}{2}(0^-)$	497.67 ± 0.13		50% K_{SHORT}	50% K_{LONG}
K^0_{SHORT}	$\frac{1}{2}(0^-)$		$\tau=(0.8923\pm0.0022)\times10^{-10}$	$\pi^+\pi^-$ $\pi^0\pi^0$	68.4 31.4
K^0_{LONG}	$\frac{1}{2}(0^-)$		$\tau=(5.183\pm0.040)\times10^{-8}$	$\pi^0\pi^0\pi^0$ $\pi^+\pi^-\pi^0$ $\pi^\pm\mu^\mp\nu$ $\pi^\pm e^\mp\nu$	21.5 12.4 27.1 38.7
$K^*(892)$	$\frac{1}{2}(1^-)$	891.8 ± 0.4	$\Gamma=50.8\pm0.9$	$K\pi$	most
$Q_1(1280)$	$\frac{1}{2}(1^+)$	1270 ± 10	$\Gamma=90\pm20$	$K\rho$ $\kappa\pi$ $K^*\pi$	42 28 16
$\kappa(1350)$	$\frac{1}{2}(0^+)$	≈1350	$\Gamma\approx250$	$K\pi$	seen
$Q_2(1400)$	$\frac{1}{2}(1^+)$	1414 ± 13	$\Gamma=180\pm10$	$K^*\pi$	most
$K^*(1430)$	$\frac{1}{2}(2^+)$	1434 ± 5	$\Gamma=100\pm10$	$K\pi$ $K^*\pi$ $K^*\pi\pi$	44.8 24.6 13.0
$L(1770)$	$\frac{1}{2}(2^-)$	1770	$\Gamma\approx200$	$K^*(1430)\pi$	most
$K^*(1780)$	$\frac{1}{2}(3^-)$	1775 ± 10	$\Gamma=140\pm20$	$K\pi\pi$	most
CHARMED, NON-STRANGE MESONS					
$D^\pm$	$\frac{1}{2}(0^-)$	1869.4 ± 0.6	$\tau=(9.1^{+2.2}_{-1.5})\times10^{-13}$	$K^0\bar{K}^0X$ eX KX	48 19 16
D^0 $\bar{D}^0$	$\frac{1}{2}(0^-)$	1864.7 ± 0.6	$\tau=(4.8^{+2.4}_{-1.5})\times10^{-13}$	$K\bar{K}^0X$ KX eX	33 44 <6
$D^{*+}(2010)$	$\frac{1}{2}(1^-)$	2010.7 ± 0.7	$\Gamma<2.0$	$D^0\pi^+$ $D^+\pi^0$	64 28

Table A.5. *Abridged particle table*

Particle	Parity P Isospin I Spin J $I(J^P)$	Mass (MeV)	Lifetime τ (seconds) Width (MeV)Γ	Decays Main mode	Fraction %
D*⁰(2010)	$\frac{1}{2}(1^-)$	2007.2 ± 2.1	$\Gamma < 5$	$D^0\pi^0$ $D^0\gamma$	55 ± 15 45 ± 15

CHARMED, STRANGE MESON

Particle	$I(J^P)$	Mass (MeV)	Lifetime/Width	Main mode	Fraction %
F⁺	$0(0^-)$	2021 ± 15	$\tau = (2.2^{+2.8}_{-1.1}) \times 10^{-13}$	$\eta\pi$ $\eta\pi\pi\pi$ $\rho\phi$	seen seen seen

BOTTOM, NON-STRANGE MESON

Particle	$I(J^P)$	Mass (MeV)	Lifetime/Width	Main mode	Fraction %
B±	$\frac{1}{2}(0^-)$	$5271 \pm ?$	?	$D e\nu$ $D_s \pi_s$	Tentative discovery
B⁰ $\overline{B^0}$	$\frac{1}{2}(0^-)$	$5274 \pm ?$	?	$D e\nu$ D_s and π_s	Tentative discovery

NON-STRANGE BARYONS

Particle	$I(J^P)$	Mass (MeV)	Lifetime/Width	Main mode	Fraction %
p	$\frac{1}{2}(\frac{1}{2}^+)$	938.2796 ± 0.0027	Stable? $\tau > 10^{31}$ (yr)		
n	$\frac{1}{2}(\frac{1}{2}^+)$	939.5731 ± 0.0027	$\tau = 925 \pm 11$	$p e^- \bar{\nu}_e$	≈ 100
N(1440)	$\frac{1}{2}(\frac{1}{2}^+)$	1400–1480	$\Gamma = 120$–350	$N\pi$ $\Delta\pi$	50–70 12–28
N(1520)	$\frac{1}{2}(\frac{3}{2}^-)$	1510–1530	$\Gamma = 100$–140	$N\pi$ $\Delta\pi$ $N\rho$	50–60 15–25 12–25
N(1532)	$\frac{1}{2}(\frac{1}{2}^-)$	1520–1560	$\Gamma = 100$–250	$N\pi$ $N\eta$	35–50 40–65
N(1650)	$\frac{1}{2}(\frac{1}{2}^-)$	1620–1680	$\Gamma = 100$–200	$N\pi$ $N\pi\pi$	55–65 ≈ 30
N(1675)	$\frac{1}{2}(\frac{5}{2}^-)$	1660–1690	$\Gamma = 120$–180	$N\pi$ $N\pi\pi$	≈ 40 ≈ 60
N(1680)	$\frac{1}{2}(\frac{5}{2}^+)$	1670–1690	$\Gamma = 110$–140	$N\pi$ $N\pi\pi$	≈ 60 ≈ 40
N(1700)	$\frac{1}{2}(\frac{3}{2}^-)$	1670–1730	$\Gamma = 70$–120	$N\pi$ $N\pi\pi$	≈ 10 ≈ 85
N(1710)	$\frac{1}{2}(\frac{1}{2}^+)$	1680–1740	$\Gamma = 90$–130	$N\pi$ $N\pi\pi$	≈ 45 ≈ 50
N(1720)	$\frac{1}{2}(\frac{3}{2}^+)$	1690–1800	$\Gamma = 125$–250	$N\pi$ $N\pi\pi$	≈ 25 ≈ 70

Table A.5. *Abridged particle table*

Particle	Parity P Isospin I Spin J $I\,(J^P)$	Mass (MeV)	Lifetime τ (seconds) Width (MeV)Γ	Decays Main mode	Fraction %
N(1990)	$\frac{1}{2}(\frac{7}{2}^+)$	1950–2050	$\Gamma = 120$–400	$N\pi$ $N\eta$	≈ 5 ≈ 3
N(2080)	$\frac{1}{2}(\frac{3}{2}^-)$	2030–2100	$\Gamma = 115$–300	$N\pi$	≈ 10
N(2190)	$\frac{1}{2}(\frac{7}{2}^-)$	2120–2230	$\Gamma = 200$–500	$N\pi$ $N\eta$	≈ 14 ≈ 2
N(2200)	$\frac{1}{2}(\frac{5}{2}^-)$	1900–2230	$\Gamma = 150$–400	$N\pi$	≈ 8
N(2220)	$\frac{1}{2}(\frac{9}{2}^+)$	2150–2300	$\Gamma = 300$–500	$N\pi$	≈ 18
N(2250)	$\frac{1}{2}(\frac{9}{2}^-)$	2130–2270	$\Gamma = 200$–500	$N\pi$	≈ 10
N(2600)	$\frac{1}{2}(\frac{11}{2}^-)$	2580–2700	$\Gamma = >300$	$N\pi$	≈ 5
N(3030)	$\frac{1}{2}(?)$	≈ 3030	$\Gamma \approx 400$	$N\pi$	seen

NON-STRANGE DELTA RESONANCES (Δ)

Particle	$I\,(J^P)$	Mass (MeV)	Width	Main mode	Fraction %
$\Delta(1232)$	$\frac{3}{2}(\frac{3}{2}^+)$	1230–1234	$\Gamma = 110$–120	$N\pi$	99
$\Delta(1600)$	$\frac{3}{2}(\frac{3}{2}^+)$	1500–1900	$\Gamma = 150$–350	$N\pi$ $N\pi\pi$	≈ 20 ≈ 80
$\Delta(1620)$	$\frac{3}{2}(\frac{1}{2}^-)$	1600–1650	$\Gamma = 120$–160	$N\pi$ $N\pi\pi$	≈ 30 ≈ 70
$\Delta(1700)$	$\frac{3}{2}(\frac{3}{2}^-)$	1630–1740	$\Gamma = 190$–300	$N\pi$ $N\pi\pi$	≈ 15 ≈ 85
$\Delta(1900)$	$\frac{3}{2}(\frac{1}{2}^-)$	1850–2000	$\Gamma = 130$–300	$N\pi$ ΣK	≈ 10 ≈ 10
$\Delta(1905)$	$\frac{3}{2}(\frac{5}{2}^+)$	1890–1920	$\Gamma = 250$–400	$N\pi$ $N\pi\pi$	≈ 10 ≈ 80
$\Delta(1910)$	$\frac{3}{2}(\frac{1}{2}^+)$	1850–1950	$\Gamma = 200$–330	$N\pi$ ΣK $N\pi\pi$	≈ 20 ≈ 15 ≈ 40
$\Delta(1920)$	$\frac{3}{2}(\frac{3}{2}^+)$	1860–2160	$\Gamma = 190$–300	$N\pi$ ΣK	≈ 20 ≈ 5
$\Delta(1930)$	$\frac{3}{2}(\frac{5}{2}^-)$	1890–1960	$\Gamma = 150$–350	$N\pi$ ΣK	≈ 15 ≈ 10
$\Delta(1950)$	$\frac{3}{2}(\frac{7}{2}^+)$	1910–1960	$\Gamma = 200$–340	$N\pi$ $N\pi\pi$	≈ 40 ≈ 60

Table A.5. *Abridged particle table*

Particle	Parity P Isospin I Spin J $I(J^P)$	Mass (MeV)	Lifetime τ (seconds) Width (MeV)Γ	Decays Main mode	Fraction %
$\Delta(2420)$	$\frac{3}{2}(\frac{11}{2}^+)$	2380–2450	$\Gamma = 300$–500	$N\pi$	≈ 10
$\Delta(2850)$	$\frac{3}{2}(?^+)$	2800–2900	$\Gamma \approx 400$	$N\pi$	seen
$\Delta(3230)$	$\frac{3}{2}(?)$	3200–3350	$\Gamma \approx 400$	$N\pi$	seen

STRANGENESS-1, LAMDA RESONANCES(Λ)

Particle	$I(J^P)$	Mass (MeV)	Width	Main mode	Fraction %
Λ	$0(\frac{1}{2}^+)$	1115.60 ± 0.05	$\tau = (2.632 \pm 0.020) \times 10^{-10}$	$p\pi^-$ $n\pi^0$	64 35
$\Lambda(1405)$	$0(\frac{1}{2}^-)$	1405 ± 5	$\Gamma = 40 \pm 10$	$\Sigma\pi$	100
$\Lambda(1520)$	$0(\frac{3}{2}^-)$	1519.4 ± 1.0	$\Gamma = 15.6 \pm 1.0$	$N\bar{K}$ $\Sigma\pi$	45 42
$\Lambda(1600)$	$0(\frac{1}{2}^+)$	1560–1700	$\Gamma = 50$–250	$N\bar{K}$ $\Sigma\pi$	≈ 25 ≈ 50
$\Lambda(1670)$	$0(\frac{1}{2}^-)$	1660–1680	$\Gamma = 25$–50	$N\bar{K}$ $\Sigma\pi$ $\Delta\eta$	≈ 20 ≈ 40 ≈ 25
$\Lambda(1690)$	$0(\frac{3}{2}^-)$	1685–1695	$\Gamma = 50$–70	$N\bar{K}$ $\Sigma\pi$	≈ 25 ≈ 30
$\Lambda(1800)$	$0(\frac{1}{2}^-)$	1720–1850	$\Gamma = 200$–400	$N\bar{K}$ $\Sigma\pi$	≈ 35 seen
$\Lambda(1800)$	$0(\frac{1}{2}^+)$	1750–1850	$\Gamma = 50$–250	$N\bar{K}$ $\Sigma\pi$	≈ 35 ≈ 30
$\Lambda(1820)$	$0(\frac{5}{2}^+)$	1815–1825	$\Gamma = 70$–90	$N\bar{K}$ $\Sigma\pi$	≈ 60 ≈ 10
$\Lambda(1830)$	$0(\frac{5}{2}^-)$	1810–1830	$\Gamma = 60$–110	$N\bar{K}$ $\Sigma\pi$	≈ 10 ≈ 55
$\Lambda(1890)$	$0(\frac{3}{2}^+)$	1850–1910	$\Gamma = 60$–200	$N\bar{K}$	≈ 30
$\Lambda(2100)$	$0(\frac{7}{2}^-)$	2090–2110	$\Gamma = 100$–250	$N\bar{K}$	≈ 30
$\Lambda(2110)$	$0(\frac{5}{2}^+)$	2090–2140	$\Gamma = 150$–250	$N\bar{K}$ $\Sigma\pi$	≈ 15 ≈ 30
$\Lambda(2350)$	$0(\frac{9}{2}^+)$	2340–2370	$\Gamma = 100$–250	$N\bar{K}$ $\Sigma\pi$	≈ 12 ≈ 10
$\Lambda(2585)$	$0(?)$	≈ 2585	$\Gamma \approx 300$	$N\bar{K}$	seen

Table A.5. *Abridged particle table*

Particle	Parity P Isospin I Spin J $I(J^P)$	Mass (MeV)	Lifetime τ (seconds) Width (MeV)Γ	Decays Main mode	Fraction %
STRANGENESS-1, SIGMA RESONANCES					
Σ^+	$1(\frac{1}{2}^+)$	1189.36 ± 0.06	$\tau = (0.800 \pm 0.004) \times 10^{-10}$	$p\pi^0$ $n\pi^+$	52 48
Σ^0	$1(\frac{1}{2}^+)$	1192.46 ± 0.08	$\tau = (5.8 \pm 1.3) \times 10^{-20}$	$\Lambda\gamma$	≈ 100
Σ^-	$1(\frac{1}{2}^+)$	1197.34 ± 0.05	$\tau = (1.482 \pm 0.011) \times 10^{-10}$	$n\pi^-$	≈ 100
$\Sigma(1385)$	$1(\frac{3}{2}^+)$	$(+)1382.3 \pm 0.4$ $(0)\ 1382.0 \pm 2.5$ $(-)1387.4 \pm 0.6$	$(+)\Gamma = 35 \pm 1$ $(0)\ \Gamma \approx 35$ $(-)\Gamma = 40 \pm 2$	$\Lambda\pi$ $\Sigma\pi$	88 12
$\Sigma(1660)$	$1(\frac{1}{2}^+)$	1630–1690	$\Gamma = 40$–200	$N\bar{K}$	≈ 20
$\Sigma(1670)$	$1(\frac{3}{2}^-)$	1665–1685	$\Gamma = 40$–80	$N\bar{K}$ $\Lambda\pi$ $\Sigma\pi$	≈ 10 ≈ 10 ≈ 50
$\Sigma(1750)$	$1(\frac{1}{2}^-)$	1730–1800	$\Gamma = 60$–160	$N\bar{K}$ $\Sigma\eta$	≈ 30 ≈ 40
$\Sigma(1775)$	$1(\frac{5}{2}^-)$	1770–1780	$\Gamma = 105$–135	$N\bar{K}$ $\Lambda\pi$ $\Lambda(1520)\pi$	≈ 40 ≈ 20 ≈ 20
$\Sigma(1915)$	$1(\frac{5}{2}^+)$	1900–1935	$\Gamma = 80$–160	$N\bar{K}$ $\Lambda\pi$	≈ 10 seen
$\Sigma(1940)$	$1(\frac{3}{2}^-)$	1900–1950	$\Gamma = 150$–300	$N\bar{K}$	< 20
$\Sigma(2030)$	$1(\frac{7}{2}^+)$	2025–2040	$\Gamma = 150$–200	$N\bar{K}$ $\Lambda\pi$ $\Sigma(1385)\pi$ $\Lambda(1520)\pi$ $\Delta(1232)\bar{K}$	≈ 20 ≈ 20 ≈ 10 ≈ 15 ≈ 15
$\Sigma(2250)$	$1(?)$	2210–2280	$\Gamma = 60$–150	$N\bar{K}$	< 10
$\Sigma(2455)$	$1(?)$	≈ 2455	$\Gamma \approx 120$	$N\bar{K}$	seen
$\Sigma(2620)$	$1(?)$	≈ 2600	$\Gamma \approx 200$	$N\bar{K}$	seen
STRANGENESS-2, CASCADE RESONANCES					
Ξ^0	$\frac{1}{2}(\frac{1}{2}^+)$	1314.9 ± 0.6	$\tau = (2.9 \pm 0.1) \times 10^{-10}$	$\Lambda\pi^0$	≈ 100
Ξ^-	$\frac{1}{2}(\frac{1}{2}^+)$	1321.32 ± 0.13	$\tau = (1.641 \pm 0.016) \times 10^{-10}$	$\Lambda\pi^-$	≈ 100

Table A.5. *Abridged particle table*

Particle	Parity P Isospin I Spin J $I(J^P)$	Mass (MeV)	Lifetime τ (seconds) Width (MeV)Γ	Decays Main mode	Fraction %
$\Xi(1530)$	$\frac{1}{2}(\frac{3}{2}^+)$	(0) 1531.8 ± 0.3 ($-$)1535.0 ± 0.6	$\Gamma=9.1\pm0.5$ $\Gamma=10.1\pm1.9$	$\Xi\pi$	≈100
$\Xi(1820)$	$\frac{1}{2}(\frac{3}{2})$	1823 ± 6	$\Gamma\approx20$	$\Lambda\bar{K}$ $\Sigma\bar{K}$ $\Xi(1530)\pi$	≈45 ≈10 ≈45
$\Xi(2030)$	$\frac{1}{2}(?)$	2024 ± 6	$\Gamma\approx16$	$\Lambda\bar{K}$ $\Sigma\bar{K}$	≈20 ≈80

STRANGENESS-3, OMEGA BARYON

Ω^-	$0(\frac{3}{2}^+)$	1672.45 ± 0.32	$\tau=(0.819\pm0.027)\times10^{-10}$	ΛK^- $\Xi^0\pi^-$ $\Xi^-\pi^0$	69 23 8

NON-STRANGE, CHARMED BARYON

Λ_c^+	$0(\frac{1}{2}^+)$	2282.2 ± 3.1	$\tau=(1.1^{+0.9}_{-0.4})\times10^{-13}$	$pK^-\pi^+$ $p\bar{K}^0$ e^+X pe^+X	≈2 ≈1 ≈5 ≈2

Indexes

NAME *and* SUBJECT

NAME INDEX

SUBJECT INDEX